TALZOYA

Of Lovers, Lonely Hearts, and the Psychotic Spell Called Falling in Love

Contents

Preface

This book is born out of an introspective effort. I am a physicist by education and consider myself to be a reasonably rational person. When it comes to my intimate life, I take an active role in selecting and engaging the mates I wish to date or be in a relationship with. For me, it is all about good sex and good companionship. I knew by my midteens that I never wanted to marry or have children. This view has not changed in the more than two-and-a-half decades that followed. It is as much an enduring characteristic of mine as the color of my eyes. I fell in love for the first—and only—time in my life several years ago. Once I came out of it—by which I mean once I entered stage four of the psychodynamic model of falling in love, which is presented in this book—analytically minded as I am, I became obsessed with wanting to understand what had happened to me. The fact that I had never regarded myself as a romantic person made this quest all the more important.

I am highly sexual and adventurous, but not particularly romantic. I have always been an avid reader of both fiction and nonfiction books, as well as peer-reviewed scientific publications. I built my model through patient self-observation and the understanding that I gained from my readings over the course of the few years that followed. My goal in writing this book is to help others who are currently in love, or who have fallen out of it, to better understand this special personal experience. More importantly, I want to help them to get a better grasp of the reasons for the life-altering decisions they may have made as a direct result of being in love. It was a delightful time of my life and resulted in tremendous personal insight and knowledge, but knowing who I am, I am infinitely grateful that being in love ended in rejection for me. Sometimes, an event we may have experienced as hurtful on the spot ends up being one of the better things that ever happened to us in retrospect. I enjoyed being in love, but I do not wish to fall in love again. I generally like being closer to reality. I like being in control of my life. I hope that this book will help my readers in some way along their own introspective journey.

Acknowledgment

I want to thank my editor, Andrew Yackira, for his diligent work and unwavering professionalism.

1

Falling in Love: Humanity's Core Shared Experience

"In the composition of the human frame, there is a great deal of flammable matter, however dormant it may lie for a time, and . . . when the torch is put to it, that which is within you may burst into a blaze." This is George Washington's description, in 1795, of what falling in love can feel like in a letter he wrote to his young step-granddaughter according to his biographer, James Thomas Flexner. Fast forwarding to modern times, here is how an anonymous contributor to an online discussion describes the experience of falling in love: "The entire world stops. Suddenly everything that ever caused you pain fades away and you are filled with a feeling of delirium. The universe around you seems to explode with happiness and you believe that you have discovered something that no one has ever experienced before. The love that you feel is unique and new, it is marvelous and unstoppable. You will be in love with this person for eternity and nothing can ever happen to stop the incredible rush." Another one adds: "I couldn't control myself around this person. I wanted to be around him all the time, daydreamed about him when we weren't together, thought everything he said and did was utterly brilliant, and for the first time in my life, thought that committing myself for life to another person wouldn't be so bad." Many admit to frequent fantasies of an intensely sexual nature about their loved one and to feeling joyful, energetic, restless, uninhibited, impulsive, and creative in the throes of love. This mental state has inspired countless ideas, paintings, songs, poems, and novels across the cultures and across the ages.

Just like people in love, those describing euphoric hypomania—a mental state commonly associated with bipolar II disorder—report having felt ecstatic, on top of the world, invincible, jumpy, energetic, hungry to take action and to accomplish goals, hypersexual, hypersocial, talkative, attractive, confident, impatient, extravagant, restless, and uninhibited. Hypomania is distinct from mania—the polar opposite of depression—in that there is no significant functional impairment. In 1930, Virginia Woolf, who is known to have had bipolar disorder, wrote the following in a letter addressed—per the chronicles of Nigel Nicolson and Joanne Trautmann—to her friend and composer Ethel Smyth: "As an experience, madness is terrific I can assure you, and not to be sniffed at; and in its lava I still find most of the things I write about. It shoots out of one everything shaped, final, not in mere driblets, as sanity does. And the six months—not three—that I lay in bed taught me a good deal about what is called oneself." Online, an anonymous blogger describes euphoric hypomania in the following way: "It vibes through the body like a wonderful buzzing summer day. Everything is enhanced—colors, feelings, possibilities, beauty, sexuality, belief in yourself, shopping, music, creativity."

The psychedelic drug lysergic acid diethylamide, commonly known as LSD, has the ability to promote feelings of happiness, peace, and a sense of becoming one with the surrounding universe. Users feel that their minds have burst through their normal boundaries. They report spiritual experiences, profound thoughts, and epiphanies that lead them to a whole new understanding of themselves and the world around them. Their perception metamorphoses. They see colorful swirls and have kaleidoscopic visions. During an LSD trip, blood flow to the control centers of the brain becomes reduced, leading to uninhibited, impulsive, and irrational behavior. Brain regions that usually function distinctly start to communicate with one another, inducing users to make unusual connections and become more creative. A good trip can feel amazing—the world can seem to explode with beauty, life, and wonder. Human interactions become deeper and more meaningful, and everyday experiences more magical. Add the stimulant cocaine to the mix, and trippers may become jubilant, energetic, restless, hypersexual, hypersocial, and extravagant. Aldous Huxley, who is best known for his book *Brave New World*, has publicly admitted to using LSD on a regular basis and is quoted as having said the following in a 1960 interview for *The Paris Review*: "It does help you to look at the world in a new way. And you come to understand very clearly the way that certain specially gifted people have seen the world. You are actually introduced into the kind of world that Van Gogh lived in, or the kind of world that Blake lived in. . . . While one is under the drug one has penetrating insights into the people around one, and also into one's own life. Many people get tremendous recalls of buried material. A process which may take six years of psychoanalysis happens in an hour—and considerably cheaper! And the experience can be very liberating and widening in other ways. . . . It's a very salutary thing to realize that the rather dull universe in which most of us spend most of our time is not the only universe there is."

If you noticed similarities in the description of the mental states discussed thus far, you are right on the money. Falling in love, euphoric hypomania, and the high induced by stimulant and psychedelic drugs involve some of the same brain circuitry and some of the same biochemistry. There is a large emotional, perceptual, cognitive, and behavioral overlap between what is oftentimes observed and reported by people who are in love, hypomanic, or high on recreational drugs. All three mental states represent a mild form of psychosis. The latter denotes a mental state in which a person's perceptions, thoughts, mood, and behavior are altered in a way that makes it difficult for him or her to tell what is real from what is not. Falling in love turns a common person into a superhuman the way that alchemists thought elixir could turn a common metal into gold. The lover's disconnection from reality is reflected in the misplaced feelings of euphoria, unfitting perception of a superlative or creative reality, delusional idealization of the loved one, obsessional thinking, hypersexuality, and, oftentimes, unseemly behavior and inappropriate language—all the hallmarks of both mania and drug-induced highs.

The word "love" is used in a variety of contexts—from loving a joke, to loving one's job, one's car, one's dog, one's children, one's friends or parents, one's long-term intimate partner, to being in love with someone one recently met. I use the expressions "falling in love" and "being in love" to refer to a very specific mental state. Of the kind of loves which I enumerated above, only being in love with someone one recently met refers to this mental state. Being in love with one's long-term intimate partner may correspond to the tail end or final stage of falling in love. Some people have never experienced the state of being in love, and never will. Some may experience it once or twice in their lifetimes. Those endowed with a more mercurial neurochemistry and are naturally more prone to strong affective states may experience it many times over the course of their lives.

Men and women have made choices of long-term mates over the course of human existence. We are the end products of the reproductively effective ones. These choices are encoded into ancestral "mental checklists." Because the sexes have traditionally occupied distinct niches—constrained and delineated by their differing biological makeups—the checklist is somewhat different for men and women. We each know these checklists without consciously knowing that we do. They contain the items which our brain's love circuitry—made up of some of the most ancient areas of our brain—constantly scans our environment for, hidden from consciousness. When we meet someone for whom this love circuitry registers enough checkmarks during a specific moment of our life, it gains permission from our prefrontal cortex, the executive and control center of our brain, to take the lead and trigger a chain reaction, releasing a cocktail of powerful biochemicals into our system. These biochemicals, in turn, change the way that we feel, act, and perceive the world.

A wonderful feat of our modern times has been the discovery and recognition—in the scientific, public, and political discourse—that both notions of sex and gender refer to continua, rather than to binary constructs. In a similar way, science is in the process of proving that all of our psychological traits come on continua where each point representing our specific version of a trait is embodied by the vast pool of specific DNA point mutations we inherited that go into determining the trait. Our genetic self is ordained at the moment of fertilization. The particular genetic code that won the reproductive lottery continues on to guide our development in utero, and then later as infants, toddlers, children, adolescents, and adults. The next chapter of this book explores how the foundation of our love style is shaped at the earliest stages of our development, while recognizing that falling in love is a process encoded in all of our brains regardless of where we fall along the biological assortment that defines our species. In a sense, it is a description of the diverse set of individuals who have the capacity to fall in love. This is the most technical chapter of the book. Those among my readers who are educated in the sciences will find it enjoyable and, hopefully, insightful. Readers who would rather skip the technicalities may safely proceed to the following chapter. It gets simpler, I promise.

Whom we are most likely to fall in love with, when, and for how long are largely determined by evolution. As much as we pride ourselves for being unique individuals, we are, in actuality, quite similar and predictable when it comes to the characteristics that we are likely to find most attractive in a potential long-term mate. We are also quite similar in how we tend to prioritize these characteristics based on an internal gauge of our own mate value and the environmental constraints we may be facing. Social scientists have put forth a great deal of effort toward uncovering what lies beneath the term "attractive" in the reproductive context. The object of the third chapter will partly be to explore what they have discovered as a result of this work. Secondly, the adage that love can strike at any time is not entirely true. There are moments in our lives where we are more likely to fall in love. The timing of this special mental state coincides with the confluence of the two drivers of evolution—namely survival and reproduction. I will delve into the details of what this means. Thirdly, the duration of this special mental state has also been largely shaped by evolution. The demands associated with the development of a human being from an embryo into an independent adult have slowly chiseled this mental state into its current form. It is our big brain that distinguishes us most from other animals. A peek into the findings of anthropology will allow us to understand how falling in love may have changed over time as our brain grew bigger and more complex and as its maturation lengthened.

In the fourth chapter, I will propose a four-stage psychodynamic model of falling in love. The first stage is a split-second rise in emotional intensity when our brain's love circuitry first starts to flood our system with the biochemical cocktail of attraction. The second stage is the core stage of falling in love. This is a stage where we slowly enter a state of mild psychosis, which can vary in duration—anywhere from two to three months to two to three years—and interestingly determined by the average time it took men and women ancestrally to pair up and achieve pregnancy. This stage consists of a linear amplification in emotional intensity over time with the goal of driving us into reproductive action. The third stage is a rapid decline in emotional intensity, again within a split second, down to a small percentage of the peak value. What triggers this stage can be rejection, achieving pregnancy, or a simple time-out in the absence of both. The fourth stage is a slow, exponential decay of the remaining emotional energy taking place over four to five years with the goal of keeping the pair-bond intact long enough to get through the pregnancy and the first few critical years of the newborn's life. I will illustrate and explain these four stages by telling the story of three imaginary protagonists caught in a love triangle.

Our brains come equipped with cortico-limbic programs that are genetically preset to run certain biochemical chain reactions within specific brain circuitry when triggered by certain environmental cues in certain environmental circumstances. Our parsimonious brains seem to co-opt the same circuitry and biochemistry to experiences as diverse as being in love, the mental states induced by psychoactive drugs such as cocaine, amphetamines, LSD, or heroin, those felt in psychosis-spectrum disorders such as mania, schizophrenia, and depression, and those that result from peak experiences as varied as achieving a professional dream, creating a piece of art, founding a company, winning a marathon, or the religious and mystical feeling of being one with the universe. Science has made great strides in understanding which hormones and neurotransmitters are involved in the attraction and attachment processes that go into forming a durable pair-bond and what each biochemical contributes to these processes. Imaging studies of the brain in love have started to map out the circuitry involved in triggering, maintaining, and terminating this mental state. The fifth chapter explores the details of the biochemistry of love and the anatomy of our brain's love circuitry.

Preexisting psychopathology and falling in love do not always mingle well together. The goal of the sixth chapter will be to explore the interplay between the two. Unrequited love is love that is not reciprocated and can sometimes have severe consequences when the lover refuses to accept defeat and move on, including a host of negative emotions and destructive behaviors such as anger, jealousy, desire for revenge, low self-esteem, rumination, hopelessness, feelings of loneliness and anguish, clinical depression, stalking, violence, and, in the most unfortunate cases, suicides or homicides. In addition, there are mental conditions that are closely tied to being in love or have their roots in the reproductive drive—namely erotomania, hypersexuality disorder, delusional jealousy, the Couvade syndrome, and false pregnancy, among others. I will go into the details of these conditions.

Unfortunately—or fortunately, depending on how you feel about it—being in love does not last a lifetime. The unshakeable belief we tend to hold while in the throes of love—that we will be in this mental state forever—is a mere delusion. Our life may well be cut short, were we to remain detached from reality for longer than necessary. If nature got its way and pregnancy has been achieved, the developing little human will need two parents who are alive and around—and sufficiently in touch with their surroundings—to bring it into the world and continue caring for it until it becomes adequately independent. In our awakening, we are sometimes lucky to find a mate whose actual self is not horribly different from the idealized mental image we projected onto them while in love. The two of us are hopefully compatible enough to make it a lasting relationship, assuming that neither finds themselves too overwhelmed by the heavy demands of parenting. Deviations from this ideal scenario are, however, frequent and the awakening can sometimes be quite rude. The last chapter will consist of a discussion of the latter topic and, more generally, of falling in love as a way of finding a long-term mate.

Whenever used, the terms "norm" or "normal" will refer to a range of values within one standard deviation of the statistical mean of bell-curve-shaped distributions. Group averages, whenever given, are never meant to apply to any single individual. The terms "outlier" or "eccentric" will denote a case that is far enough from the mean or average, typically two or more standard deviations away. This is in recognition of the fact that any of our traits, whether physical or psychological, come on continuous spectra. The diversity among us is what makes humanity the resilient and strong species it is. Being at one end of a spectrum or the other is neither good nor bad in and of itself. The benefit or harm of being one way rather than the opposite can only be sized up relative to the specific circumstances confronting the individual. The expressions "mental illness" or "psychopathology" are used to refer to a pathological mental state that results when an enduring group of psychological traits an individual possesses causes significant impairment in the individual's day-to-day functioning if faced with certain environmental stresses. Those of us who are endowed with the more eccentric versions of certain traits are oftentimes more environment-selective than those possessing middling versions of the same traits. This is the reason why mental illness is typically intermittent, with long periods of time where the affected individual functions quite well—at what would be regarded a "normal" or "better" level—interrupted by periods of distress.

Being in love is being in a mildly psychotic mental state in the sense of becoming somewhat detached from reality or not always being able to tell what is real from what is fantasy. Being in a mild state of psychosis is not necessarily a clinical event. The proof is that none of us typically seeks psychiatric treatment to cure us from being in love, unless another psychopathology is also at play. The reason for this is that it is a peak experience, the highest level of delight a human being can ever experience in a lifetime. It is a delight that, at times, can have disastrous consequences—which may occasionally require psychiatric intervention—but it can also lead to wonderful outcomes, one of which is the birth of new human being. Many with bipolar II disorder, whose moods oscillate between euphoric hypomania and clinical depression, are often misdiagnosed as unipolar depressive because they rarely ever seek treatment for the hypomanic episodes. It is the depression that really bothers them. Hypomania, on the other hand, can confer an enormous advantage both socially and professionally. Creativity almost always requires some level of detachment from our surroundings or, stated otherwise, becoming somewhat psychotic. I doubt that anybody will want a cure from being creative—unless, perhaps, if the creativity is of the destructive kind. But then again, even creative destruction can sometimes have beneficial outcomes. My goal in writing this book is to stimulate thoughts and entertain. I hope that what follows makes for instructive and enjoyable reading.

2

Fetal Development and How It Lays the Foundation of Our Love Style

On any given day, even if mundane, our smooth functioning relies on countless small decisions made unconsciously by our minds. Indeed, given the complexity of the world that surrounds us, the multitude of possible choices in any of the innumerable situations we may encounter, and faced with time and processing power limitations, our brain functions on autopilot most of the time. It does so for the sake of efficiency, by way of heuristics or the use of simple rules of thumb and approximations. Daniel Kahneman calls this "thinking fast," as opposed to the "slow thinking" we do when we consciously engage our prefrontal cortex to perform a deliberate analysis of our circumstances before deriving a conclusion. We use mental shortcuts because we get it right more often than wrong when we do it systematically. For instance, if a rule is true seven times out of ten, and we use it in every single situation where it is appropriate, we will statistically get it right more often than not. Indeed, applied enough times, we will be right about seven times out of ten. We will also be wrong three times out of ten—but that is good enough for our overwhelmed brain. One example of this process is our unconscious use of stereotyping, or the blind projection of group averages onto single individuals. In order to choose the heuristic most appropriate for a given situation, our brain relies on pattern matching and schemas, which are clusters of words, images, emotions, and ideas mentally connected to each other by links of varying strengths. Thus, when I say "female," I activate a schema in your mind which may include items such as long hair, a dress, a bra, makeup, the chromosomal sex XX, breasts, the image of a vagina or that of a uterus, the idea of menstrual cycles, pregnancy, the traditionally female-typical role of caring for vulnerable individuals, a familiar smile, the idea of being more friendly or maybe more moody, the female-typical tendency to put emotional intimacy ahead of sexual intimacy, to be more people-oriented rather than object-oriented, what it felt like to break up with your first girlfriend, or the face of your mother or a childhood friend. Of all the elements in this cluster, our brain picks one or more most appropriately matching the pattern it detected on the spot and makes us feel or think or act a certain way. We inherit many of these schemas in their skeletal forms from the long evolutionary history of humanity and flesh them out to a large extent during our childhood, and to a lesser extent later in life. Each of us has inherited a slightly different version of the schemas and, sometimes, unique schemas. We all add a few schemas of our own during our lifetime and sometimes pass some of them down to our children.

Our stereotypical way of thinking of the human male and female tends to be very binary. The sexes seem to stand at distant opposite poles, each endowed with its own set of stereotypical attributes, and nothing but a long stretch of void separating the two. It is either one or the other, much like either black or white. But a great feat of modern times has been the recognition, in both the scientific and public circles, of the fluidity of the concepts of both sex and gender. The fact that it should have taken us this long to acknowledge their continuous nature is testimony enough that doing so is not trivial. Many of us stand in the grey zone that connects the stereotypical poles, not just in our likes or dislikes, but also in our personality, abilities, proclivities, and our anatomy—both external and internal.

Simon Baron-Cohen identifies five levels of sex: (1) the *genetic sex*, which refers to the (46, XY) genotype typical of a male and the (46, XX) genotype typical of a female; (2) the *gonadal sex*, which refers to the presence of two testes in a typical male and two ovaries in a typical female; (3) the *genital sex*, meaning a penis and scrotum for a normal male and a clitoris and vagina for a normal female; (4) the *brain type*, referring to the tendency of a typical male brain to be more systemizing (meaning system-centric, object- and task-oriented) than emphasizing (meaning people-centric, care- and relationship-oriented) and that of a typical female brain to be more emphasizing than systemizing; (5) the *sex-typical behavior*, which refers to men's predilection for cars, team sports, mechanics, statistics, stamp collections, and activity-driven social groups and women's penchant for fashion, cosmetics, gossiping about the promiscuous sexual behavior of rivals, and intimate one-on-one relationships.

When we talk about falling in love, matters get even more complicated. David Buss identifies three senses of the phrase "sexual orientation": (a) the *primary sexual orientation*, which refers to whom one is attracted (men, women, or both); (b) *gender identity*, which refers to whether one subjectively feels like a woman, a man, both, or neither; and (c) *sexual behavior*, which refers to the gender of the individuals with whom one actually has sex.

Within the context of intimate relationships, another important dimension on which people can differ quite a bit is libido or sexual appetite, which corresponds to the level of an individual's general interest in being sexual or in engaging in sexual activity. Men, on average, tend to have a stronger libido than women. But the distributions of sexual interest for the two sexes overlap. And the differences between the individuals within a given sex are much bigger than the gap between the mean of the aforementioned distributions. Any given individual could have a strong libido, an average libido, a low libido, or a libido so low the individual may be classified as asexual. Libido has some correlation to physical sexual function or dysfunction. Did you know, for instance, that roughly 43% of women and 31% of men ages 18 to 59 in the US (Laumann et al., 1999) have some form of sexual dysfunction? For women, having a low libido dominates among the possible dysfunctions, but about 10% of women suffer from endometriosis (in which the endometrium—the tissue lining the uterus—spreads to extrauterine tissue and often results in painful intercourse), and 5% to 17% of women may suffer from vaginismus. This last condition is thought be a form of anxiety disorder which leads to such strong contraction of the vaginal muscles as to make penetration—by anything from a penis to a tampon—impossible. The actual percentage for vaginismus is not known, because many women seem to be too embarrassed to discuss it even with their primary physician. Vaginal atrophy, another prevalent condition which can make intercourse quite painful and about which very little tends to be known in the general population, is very common in perimenopause and menopause, but can also happen postpartum to lactating younger women. Again, embarrassment seems to drive the general state of ignorance. Up to 50% of postmenopausal women may have it. The major dysfunctions for men are erectile dysfunction, premature ejaculation, and low libido. For erectile dysfunction, the rule of thumb to figure out the prevalence rate is to roughly add 10% per decade of life. The frustration of dealing with physical dysfunctions tends to have a depressing effect on libido so that the affected individual may end up losing nearly all interest in sex over time.

In our stereotypical representation of a female, the aforementioned five levels of sex typical of a female are aligned. In addition, we view her as being attracted to males, feeling like a female, actually having sex with men exclusively, and we do not even think for a moment that she could have any sort of sexual dysfunction. And in our stereotypical representation of a male, all the five levels of sex typical of a male are aligned. In addition, he is attracted to females, feels like a male, has sex exclusively with females, and certainly does not manifest any sort of sexual dysfunction. In reality, many combinations and permutations of the five levels of sex and the three senses of sexual orientation are observed. Libido could be anywhere on the scale and physical sexual dysfunction can be a part of the mix. The reason that I am emphasizing this point is that it will become very pertinent to the discussion in the last chapter of this book about the rude awakening from the intoxicating effect of falling in love. The process of falling love that we have evolved as humans, a process by which our brain attempts to rapidly transform a total stranger into kin, is entirely blind to this majorly important detail.

Much of the above differences between individuals are forged in the womb.

The genetic sex is the first step of sexual differentiation. You may remember from your secondary school biology classes that we each have twenty-three pairs of chromosomes. Each pair contains a chromosome we inherited from our mother and a chromosome given to us by our father. Each oocyte (or immature egg cell) in the ovary contains a preset sample of twenty-three chromosomes, each randomly selected from our mother's twenty-three pairs. Similarly, each spermatozoon (or motile sperm cell) contains twenty-three chromosomes which have each been randomly chosen from the twenty-three pairs carried by our father. If sexual intercourse results in fertilization, the winning ovum (or mature egg cell) is penetrated by a spermatozoon to form a zygote, a single cell which now has a complete set of forty-six chromosomes and is destined to develop into an embryo, which later—at around the twelfth week of gestation—will be called a fetus. These forty-six chromosomes represent the complete genetic program we call our DNA, the blueprint responsible for our pre- and postnatal development and every single one of our traits, whether physical or psychological. One of these twenty-three pairs consists of the sex chromosomes. Our mother had the XX form of that pair and gave us one of her two X chromosomes. So, it is really the sex chromosome carried by the spermatozoon that determines our genetic sex: the zygote is genetically male if it contains the father's Y chromosome and female if it has the father's X chromosome. This seems simple enough: the genotype of the typical female is (46, XX), and the genotype of the typical male (47, XY). But nature likes diversity, or sometimes—a bit like us—it just makes a mistake, the result of which is that genetic sex is not clear-cut. Women with Turner syndrome have the (45, X) genotype and cannot get pregnant due to early loss of ovarian function. This is a severe condition typically resulting in short stature and skeletal abnormalities noticeable from an early age. Those with the triple X syndrome have the (47, XXX) genotype and may be fertile. Most women with this condition are never diagnosed because it does not usually result in any divergence from the norm worthy of notice besides a taller-than-average stature. A man with Klinefelter syndrome has the (47, XXY) genotype and is sterile unless medically assisted. For men with this genotype, the symptoms could be severe, resulting in the early diagnosis of the condition, or they may be so mild that it may not get diagnosed until puberty or even adulthood, if it gets diagnosed at all.

In the first few weeks of development, assuming that the embryo has either the typical male or the typical female genotype, three primordial structures can be discerned which will eventually differentiate sexually: the Müllerian (or female) structure, the Wolffian (or male) structure, and two undifferentiated (or hermaphrodite) gonads. If the Y chromosome is present, the gonads normally start to differentiate into testes by the sixth week of gestation. If not, they remain undifferentiated until the twelfth week following conception, at which time they start evolving into ovaries. This is the stage where our second level of sex, the gonadal sex, is set.

By the seventh week of pregnancy, the differentiated testes start producing hormones. They secrete two types of hormones: the anti-Müllerian hormone (AMH) and androgens (for instance, testosterone). In the absence of these hormones, the embryo develops according to the default template, which is female, presumably under the influence of the mother's female hormones (most likely estrogen). Between roughly the eighth and the twelfth weeks (or during the third month) of gestation, internal genitalia differentiate. The presence of AMH causes the regression of the Müllerian structure which would otherwise develop into fallopian tubes, a uterus, a cervix, and upper one-third of the vagina. Meanwhile the presence of androgen allows the Wolffian structure to develop into the duct system leaving the testes—including the epididymis, vas deferens, seminal vesicle, and ejaculatory duct. In the absence of androgen, the Wolffian structure spontaneously regresses. Since AMH is also absent in females at this stage, the Müllerian structure is allowed to develop into the normal internal genitalia of a female—fallopian tubes, a uterus, a cervix, and upper vagina. Deficiency of AMH in a (46, XY) genetic male leads to a condition known as persistent Müllerian duct syndrome (PMDS) characterized by undescended testes and the presence of an underdeveloped uterus and fallopian tubes in an otherwise normal-looking male. The deficiency of AMH may be caused by a mutation in the gene located on chromosome 19 which codes for it, or that for the AMH receptor, but could also result from insensitivity of the AMH receptors on the target organ. A PMDS male can experience fertility issues.

Using the expression "male hormone" to refer to testosterone or to AMH and "female hormone" to refer to estrogen is quite misleading in the sense that all three hormones are produced in both men and women. The word testosterone is derived from the word "testes" and estrogen from the word "estrus" (the moment a female animal is "in heat"), which naturally leads us to label them as "male" versus "female." And then there is the fact that males, on average, produce more testosterone than females, and females more estrogen than males. We tend to forget that the distribution of the levels of each of these hormones for men and women overlap. The term AMH is originally devised to refer to the inhibiting effect it has on the development of the Müllerian structure for a male embryo, but it is also produced in the ovaries of the female. It differs in the timing of production and the function it plays in the male versus the female.

AMH is produced by the Sertoli cells in the testes of the male embryo and plays a central role in the development of the normal internal sexual organs of the male by promoting the regression of the original female structures (as described above). But AMH is also produced by the granulosa cells that surround the oocyte inside each follicle (or sac) in the ovaries of a female. The granulosa cells envelop each egg and provide it with energy. The AMH they produce serves to inhibit excessive recruitment of follicles by the follicle-stimulating hormone (FSH) released by the brain in the process of selecting the dominant follicle for ovulation. Their number naturally diminishes over time as follicles are recruited for ovulation or die during the selection of the dominant one every month, and with it, the levels of AMH. Low blood level of AMH in the adult female is, in fact, the best predictor of the onset of irregular periods, a marker of the start of perimenopause. Blood levels of AMH in a female are low at birth, grow slowly until puberty, remain roughly stable during the core of her reproductive years, start to decrease with the approach of menopause, and become essentially undetectable at menopause. Additionally, in a condition called polycystic ovary syndrome (PCOS), cysts—or enlarged semimature follicles which have insufficiently developed from immature follicles due to disturbed ovarian function and have therefore failed to either die or ovulate—remain present in the enlarged ovaries and account for the abnormally high number of granulosa cells, hence excessive blood level of AMH. Therefore, unusually high blood level of AMH is a good indicator of PCOS. This condition is another type of female sexual dysfunction and can lead to elevated androgens in females, delayed puberty, irregular or absent periods, heavy periods, excess facial and body hair, pelvic pain, and fertility issues.

It is commonly known that testosterone is produced by the testes in males, and estrogen by the ovaries in females. What is often overlooked is the fact that our adrenal glands are majorly involved in the production of sex steroids as well. The adrenal glands are triangular-shaped endocrine glands sitting on top of each of our kidneys. They have one of the greatest rates of blood supply per gram of tissue of any organ, which is testimony to the crucial role that they play in regulating the human body. The adrenal glands of a human fetus are first detectable around the sixth week of development. By the eighth week of gestation, they show steroid hormone production capabilities (Ishimoto and Jaffe, 2011). The androgens they generate play a central role in shaping the brain, hence the postnatal behavior, of a female fetus. All brains are female until roughly the eighth week of gestation, at which point elements of the male-typical brain start to be etched into them. I will come back to the very important topic of fetal brain development and its impact on later male-typical or female-typical adult behavior. While the differentiation of the internal genitalia takes place between roughly the eighth and twelfth weeks (or third month) of gestation, the differentiation of the external genitalia starts around the same time period and extends into the third trimester of gestation. Development of the male external genitalia (a penis and scrotum) requires the presence of dihydrotestosterone, a hormone derived from testosterone, the default being female external genitalia (a clitoris and labia). In adulthood, the level of testosterone is central to setting the level of the libido in both the male and the female, a topic that I will also further delve into later. I want first to open a parenthesis on the general role played by our adrenal glands.

Each of our adrenal glands has an outer layer, called a "cortex," and an inner area, called a "medulla." Each of these two areas produces its own set of hormones. The adrenal cortex is responsible for the production of three main types of steroid hormones: mineralocorticoids, glucocorticoids, and androgens. The most important mineralocorticoid to know about is aldosterone, which helps to regulate salt (sodium and other electrolyte) levels in the blood. More precisely, the level of aldosterone set by our adrenal glands dictates the amount of salt and water that our kidneys should be discharging into urine versus the amount that should be left to circulate in the blood. This, in turn, influences our blood pressure. The glucocorticoids produced by the adrenal cortex, cortisol and cortisone, play important roles in the regulation of metabolism and immune system suppression. Cortisol regulates the amount of sugar circulating in the blood by helping the body to free up the necessary ingredients from storage (fat and muscle) to make glucose (a simple sugar). The glucose is converted to energy to power up our cells during the fight-or-flight response, also called the "stress response." The immune system is suppressed because energy is preferentially allocated for peak activity in the presence of an imminent threat to our survival—historically, a wild animal or someone from an enemy tribe for our ancestors. The main androgens produced by the adrenal cortex are dehydroepiandrosterone (DHEA), DHEA-sulfate (DHEAS), and in far lesser amounts, androstenedione, testosterone, and dihydrotestosterone. The adrenal glands also release small amounts of estrogen. DHEA is a precursor hormone, derived from cholesterol, and is believed not to be potent by itself. Instead, it becomes potent when converted into active sex steroid hormones, such as testosterone and estradiol (a very potent type of estrogen). Some of that conversion happens in the adrenal cortex itself. In fact, both the DHEAS and androstenedione are derived from DHEA in the adrenals. The DHEAS is released into the bloodstream and accounts for almost all of the circulating DHEA in the human body. While some of the androstenedione is directly released into the bloodstream, some of it can be further converted into testosterone or estrone (another type of estrogen) within the adrenal glands before being released into the bloodstream. Some of the testosterone and some of the estrone can further be aromatized into estradiol before being released. About 10%

of testosterone is usually converted to dihydrotestosterone (a more potent form of the hormone). Recent research reveals that peripheral tissue—such as brain, liver, skin, bone, or muscle—can synthesize active sex steroids from circulating DHEAS for local use. They do that by first turning DHEAS back into DHEA, and then triggering the cascade of reactions described above. DHEA is also synthesized in ovarian theca cells, testicular Leydig cells, and de novo in the brain via conversion of cholesterol. The adrenal medulla, the inner zone of the gland surrounded by the cortex, produces the catecholamines epinephrine (also called *adrenaline*) and norepinephrine (also called *noradrenaline*). Epinephrine and norepinephrine are involved in the fight-or-flight response, characterized by a quickening of breathing and heart rate and an increase in blood pressure via constriction of blood vessels. The adrenal glands of the fetus are much larger in relation to its overall body size than those in adults. At twelve weeks of gestation, for instance, the adrenal glands are four times the size of the kidneys. The fetal adrenal cortex is also different from its adult counterpart in that it is composed of two zones, the fetal zone and the definitive zone which is still developing. The medulla does not develop much until after birth. The fetal zone of the cortex produces large amounts of adrenal androgens. We are talking about fetal DHEA and DHEAS levels that are sometimes tenfold greater than the levels measured in adults! A large fraction of the DHEAS in circulation is turned into estradiol by the maternal placenta and helps to regulate pregnancy. Adrenally produced cortisol is essential for prenatal development of organs, particularly for the maturation of the lungs. The fetal zone of the adrenal cortex rapidly disappears after birth, resulting in decreased size of the adrenal glands and decreased androgen production. Cortical androgen production remains low in early childhood until just before puberty, a time period called adrenarche, characterized by increased DHEA and DHEAS production. Increased levels of adrenal androgens are responsible for the development of axillary and pubic hair before the beginning of puberty in both boys and girls.

The understanding of the sexual differentiation of the brain is a scientific work in progress. We know that the default template is female, and the presence of androgens is needed to impart elements of the male-typical brain. Circulating testosterone is locally converted into estradiol when it reaches the androgen receptors of the brain by a process called aromatization. This estradiol, in turn, activates neural estrogen receptors to induce masculine-type development (McCarthy, 2008). In the preceding paragraph, I mentioned the very high circulating concentrations of DHEAS in the womb and talked about how this precursor hormone can be turned into testosterone and estradiol upon reaching the brain. Thus, circulating adrenal DHEAS could have the same masculinizing effect in the brain of a female fetus as does gonadal testosterone in the brain of the male fetus. Remember, a large fraction of the DHEAS produced by the fetal adrenal cortex is converted into estradiol by the placenta. In fact, maternal levels of estradiol are so high during gestation that circulating estradiol, if it were to reach the fetal brain, would entirely obscure the effect of estradiol aromatized from circulating testosterone in the brain and nullify any possibility of sexual brain differentiation by action of this hormone. It turns out that a binding globulin present in the bloodstream (called *alpha-fetoprotein*) with a high affinity for estradiol sequesters it in the bloodstream by latching onto it, and thus prevents it from masculinizing the brain. The testes of males produce increasingly high levels of testosterone from the eighth to the twenty-fourth weeks of gestation. Some amount of testosterone is also secreted by the oversized fetal adrenal cortex starting with the eighth week following conception and by the ovaries of the female fetus shortly after their differentiation in the twelfth week of gestation. Testosterone and estrone can also be derived in the brain from the very high concentrations of adrenal DHEAS in the fetus. Estradiol, whether it is aromatized from all this testosterone and estrone or even made de novo locally in the brain from cholesterol, masculinizes it through a process called cerebral lateralization (Geschwind and Galaburda, 1985). It affects the rate of growth of the two hemispheres by increasing the rate of growth of the right hemisphere and slowing down the rate of growth of the left hemisphere. The right hemisphere is where the spatial modules that underly a lot of the abilities involved in success with the male-dominated STEM (Science, Technology, Engineering, and Mathematics) fields reside. The male-typical brain has larger modules for the processing of both visual and sexual stimuli. The right hemisphere is, on average, bigger in the male brain. The left hemisphere is where language modules reside. In most males, those modules are solely present in the left hemisphere, while they seem to be duplicated in the right hemisphere of most females. This consequently reduces the area available for the processing of spatial information in the right hemisphere of the typical female brain. This, in turn, coupled with the fact that the left hemisphere develops faster and grows to be larger in females, accounts for the superior verbal abilities and inferior spatial and mechanical abilities observed for females, on average. Baron-Cohen generalizes these ideas by proposing that the female-typical brain is superior at emphasizing (which includes language abilities, theory of mind—meaning the ability to infer the intentions of others, and the propensity to care) and the male-typical brain is better at systemizing (meaning the ability to uncover the cause-and-effect relationships that govern a system in order to control it, and includes the kind of abilities involved in success with STEM subjects). In addition, males, on average, would be expected to be faster at processing and responding to sexual stimuli. Having a larger area to process visual information, they would be especially sensitive to sexual stimuli that are visual in nature. This may explain, at least in part, why men tend to find exposed female body parts so erotic, while women, on average, tend not to get as aroused by the sight of exposed male body parts. Group averages do not, however, tell much about individual men or women. Many men choose to join care-related professions, while many women excel in STEM fields. And I personally know a few women who would very much welcome any young man who could make the cut for Thunder from Down Under or the *Playgirl* centerfold to expose as much of his body as he wishes.

The androgens DHEA and DHEAS are together the most abundant steroid hormones in the human body. As I mentioned earlier, the levels of these hormones soar during fetal development. In adults, their concentrations steadily decline with age, becoming nearly undetectable around the time that many diseases of aging become more prevalent. This has led scientists to speculate that they may play a role in the genesis and protection of neurons. They may play an especially important role in women's health.

In the female body, half of testosterone is synthesized from circulating DHEAS in peripheral tissues—the other half being made in the adrenal glands and the ovaries, while 75% of premenopausal estrogen and 100% of postmenopausal estrogen are synthesized from DHEA. Studies on aggression, typically expressed by action of gonadal testosterone on certain brain circuitry during the mating season in the male of mammalian and avian species, indicate that aggression displayed during the nonmating season (where gonadal testosterone secretion is low) may instead be mediated by conversion of DHEA into sex steroids locally in brain circuitry (Soma et al., 2015). This gives some support to the idea that DHEA could play a role in masculinizing the brain of a female fetus, an idea I developed in the previous paragraph. In the adult female, it could be relevant to setting the level of libido via conversion to testosterone. DHEAS converted into estradiol on female genitalia could help to maintain blood flow to those regions and prevent atrophy with the approach of menopause. Testosterone is thought to have a protective effect against depression. DHEA could play the same role via conversion to testosterone. Therefore, women with the least amount of adrenal androgens may be especially vulnerable to becoming depressed. Given all of this information, one can speculate that women with moderately high adrenal androgen throughput would have a more male-typical brain and tend to display more male-typical interests and abilities (translating into more success in STEM fields), to have higher interest in sex throughout life, to be more responsive to visual sexual stimuli, to show more aggressive and competitive behaviors, and perhaps even to be less likely to experience genital atrophy, depression, or neurodegenerative diseases in older age. They would be perfectly attractive looking women with fully functional, female-typical internal and external genitalia, yet equipped with a male-typical brain, who, as a result, would thrive in the more competitive fields, be more sexually active and adventurous, display high levels of independence, and even have a short-term mating preference. Again, I personally know a few women who would perfectly fit this description. DHEA would work very much the same way for males, which may not be an issue at average or below average adrenal throughput, since its effect would be largely overshadowed by that of gonadal testosterone.

As is oftentimes the case, too much of a good thing can be a bad thing. A condition known as congenital adrenal hyperplasia (CAH) serves to illustrate what can go wrong when adrenal androgen production is on overdrive. This condition is due to a recessive genetic abnormality, meaning both parents have to be carriers of the defective gene for their child to be affected. It is associated with an inability or a deficiency in the ability of the adrenal glands to produce cortisol. Since the brain never gets a negative feedback from the detection of sufficient levels of cortisol in the blood, it unremittingly stimulates the adrenal glands into pumping out more hormones, which leads to hyperplasia (an increase in the number of hormone-producing cells in that organ). This translates into excessive production of adrenal androgens, particularly DHEA and DHEAS. While only cortisol production is lacking in 25% of the disorder's classic form, in 75% of cases the adrenals also lack the ability to produce aldosterone. Remember, aldosterone helps to regulate salt levels in the blood. Improper regulation can lead to fatal electrolyte and water imbalances. In both cases, the newborn gets hormone replacement therapy shortly after birth. Since cortisol balance is thus restored, the brain ceases to overstimulate the adrenals so that androgen levels are also normalized. Excessive levels of androgens were, however, present in the womb. In girls, this often results in prenatal virilization with ambiguous external genitalia, which exhibit both female-typical and male-typical features. This usually requires corrective surgery in the first year of life. If left untreated post-birth in boys, severe CAH can result in hypermasculinization with premature onset of puberty. Boys as young as two or three years of age can acquire all the secondary sexual characteristics that come with puberty, including deepening of the voice, growth of facial and bodily hair, acne, and excessive muscle mass, the latter leading to an abnormally short stature. The nonclassical form of CAH is both more common and milder, and may not become evident until later in childhood or early adulthood: cortisol deficiency is at 10% to 75% of the normal range, and the androgen overproduction is accordingly less severe. In all cases, with hormone replacement (and corrective surgery in the case of girls born with ambiguous genitalia), these individuals can lead a relatively average life. Excess adrenal androgen during fetal development, however, leaves CAH females with another permanent mark: a male-typical brain and the associated male-typical or "tomboyish" behaviors and preferences. As girls, they tend to enjoy energetic and outdoorsy activities, team sports, utilitarian grooming, and functional clothing. They tend to find little interest in dolls, avoid babysitting, and disdain traditionally feminine domestic tasks. They have above-average school achievement and above-average spatial skills. They reach puberty later. According to New Zealand American psychologist and sexologist John Money, they also reach sexual maturity later, starting the dating and romantic stage three to nine years later than their age-mates. They put their career ahead of marriage or motherhood. Many do get married eventually, though some have fertility issues. This supports the idea that adrenal androgens may play a central role in imparting aspects of the male-typical brain into a female's brain. What is interesting is the fact that CAH boys tend to score lower on tests of spatial and mechanical abilities. This bolsters the theory that optimal development of spatial abilities may require a moderate level of androgen exposure in fetal development, somewhere in the low-to-medium male range. The combination of high-adrenal and high-testicular androgen could also result in suboptimally developed language abilities and theory of mind, which can cause impaired social functioning in boys and men.

Another rare disorder called androgen insensitivity syndrome (AIS) is also worth mentioning in order to illustrate how the five levels of sex I introduced earlier in this chapter are not always aligned the way they tend to be in our stereotypical mental representation of the sexes. AIS is due to a genetic defect on the X chromosome in genetically male—(46, XY)—individuals. Androgen insensitivity can be complete or partial. Because of the presence of the Y chromosome, gonads normally differentiate into testes at six weeks of gestation, and they produce normal levels of both AMH (or anti-Müllerian hormone) and testosterone. Because their cells are normally sensitive to AMH, the Müllerian ducts regress, resulting in the absence of a uterus and fallopian tubes. The problem in the severe case of AIS—with complete androgen insensitivity—is that cells are not sensitive to testosterone, which is needed for both the masculinization of the brain and that of external genitalia. As a result, the affected individuals are endowed with a perfectly female-presenting body and a female-typical brain, leading to female-typical behaviors and preferences. They are given female names at birth and reared as females. It is generally only when they reach puberty and develop breasts normally but fail to menstruate (for lack of a uterus) that their condition is diagnosed. In the case of partial androgen insensitivity syndrome (PAIS), ambiguous external genitalia usually signal the condition earlier. One can speculate that a genetically male individual with moderately low levels of androgen synthesis or a relatively mild case of androgen insensitivity may have a perfectly average male-typical appearance, the internal and external genitalia of a normal male, but a female-typical brain. He would thus display high levels of empathy, disdain rough-and-tumble play in childhood, have a preference for one-on-one close relationships, enjoy spending time with children as an adult, and might even opt for a care-oriented profession. Most of us have probably encountered at least a few men who would fit this description.

I have now reached a good point to start spending some time delving into the three senses of "sexual orientation" I introduced earlier—namely, primary sexual orientation, gender identity, and sexual behavior. Testosterone is notably higher in males than in females during three periods of human development: from about weeks eight to twenty-four of gestation, during the first five months after birth, and during puberty. I discussed in detail how the behavior of CAH girls and women appears male-typical due to prenatal exposure to high levels of adrenal androgens. Studies done cross-nationally show that the majority of women with CAH are heterosexual, but about 30% are not, a figure markedly higher than the roughly 5% or fewer in the general population (Hines et al., 2015). The 30% rate is especially high given the observed rate of homosexuality in women. Most studies peg occurrence at 1% to 2% for women and at 2% to 4% for men. The fact that 70% of women with CAH show at least some heterosexual inclination, however, hints at more than just high androgen exposure in determining primary sexual orientation, though the effect of the less severe, nonclassical form, could account for it to some degree. Individuals with AIS—who are male for genetic sex and male for gonadal sex—almost always have a sexual orientation toward men, consistent with their prenatal androgen insensitivity. The anterior hypothalamus of the brain is thought to be involved in regulating human sexual orientation. A study among three groups (women, men who were presumed to be heterosexual, and homosexual men) involving measurement of the postmortem size of four cell groups in this brain region—interstitial nuclei of the anterior hypothalamus (INAH) 1, 2, 3, and 4—found the size of INAH 3 to be twice as large in heterosexual men compared to both women and homosexual men (LeVay, 1991). The size of INAH 3 is presumably established early in life, perhaps via androgen exposure during a critical window, and later influences a person's primary sexual orientation. With regard to gender identity, or one's sense of self as "male" or "female," exposure to high levels of androgens prenatally has been linked to a person's increased likelihood of developing a male-typical gender identity, despite being reared as a female (Hines et al., 2015). Cross-nationally, a change to living as a man has been observed in about 1% to 3% of women with CAH, as well as in women with other genetic conditions causing exposure to elevated concentrations of androgens during early development. Even when they do not wish to change their lives to live more fully as males, the strength of identification as "female" also seems to be reduced in women and girls with CAH. If given the choice, many would have preferred to be born male (Hines et al., 2015). Research also indicates that individuals with AIS almost always develop a female gender identity, consistent with their lack of early androgen sensitivity. This suggests that early androgen exposure could play a major role in the development of gender dysphoria later in life for people labeled female at birth, while insufficient androgen exposure during a critical window in fetal development could have the same effect on people labeled male at birth. Sexual behavior, which refers to whom one actually has sex with, seems to be the most amenable among the three senses of sexual orientation to environmental influences. Most of us, even when our primary sexual orientation is strongly heterosexual, are capable of at least some degree of homosexual behavior. The threshold to trigger such behavior is higher in some of us than it is in others. It is a historical fact that, where young men and women were physically segregated for extended periods of time (such as in times of war or in strictly sex-segregated schools, professions, and roles) homosexual behavior occurred more commonly than under modern circumstances in today's western world. Most of these men and women later went on to marry an opposite-sex individual and have children. The same is commonly observed today (although kept in strict secrecy) in those parts of the world where premarital sex is not just frowned upon, but severely punished. When the socioeconomic constraints we face are so severe that the only two options seem to be marriage and kids or social ostracism and starvation, most of us seem to have a strong enough survival instinct to pick the former rather than the latter, regardless of our primary sexual orientation or gender identity. Gaining the freedom to express the latter two openly seems to be a luxury of modern, industrialized, and well-to-do societies. Varying degrees of bisexuality may have been evolved to allow for adaptation to harsh ecologies. When the going got tough enough, having a special connection with a same-sex person could well have been lifesaving. When a trait is allowed to vary on a continuum, a small percentage of individuals are bound to end up at the extreme ends of that continuum, exhibiting exclusively heterosexual or exclusively homosexual tendencies in the case of primary sexual orientation. In that sense, homosexuality is bound to exist and to persist. The statistical fact that more men than women identify as homosexual may be an indication that having at least some level of homosexual tendency has ancestrally been more lifesaving for men than for women, on average. After all, ancestral women mainly depended on men for their survival, and oftentimes, ancestral men depended on other men as well, albeit in a different way.

The topic of sex drive or libido is central to any discussion on the subject of falling in love. Indeed, the instinctual drive to engage in sexual activity is a sine qua non for reproduction. Increasingly, research points to testosterone and the precursor androgen DHEA in determining the strength of that drive for both men and women. I mentioned previously that differential prenatal exposure to these androgens induces a dimorphism in the size of the sexual processing areas of the brain for a male versus a female fetus, making it larger in the former. This may be part of the reason why men, on average, tend to exhibit higher libido than women. Another part may be played by circulating levels of these hormones in the adult. Many studies point to androgen insufficiency as a major cause of decreased libido and urogenital atrophy in postmenopausal women. The gradual decrease of testosterone levels with age leads to a parallel decrease in the strength of the sex drive in men, as well as to physical (erectile) dysfunction, especially as the level plummets below a seemingly critical threshold around age fifty. In general, low sex drive appears to have a strong correlation with both poor physical health (as manifested in serious conditions such as obesity, high blood pressure, high cholesterol, etc.) and poor mental health (such as depression and anxiety). The direction of the cause-and-effect relationship, however, is not clearly established. Could androgen insufficiency (both pre- and postnatal) be, at least in part, responsible for both? A recent study (Guay et al., 2004) concludes that premenopausal women with complaints of sexual dysfunction had lower adrenal androgen precursors and testosterone than age-matched control women without such complaints. Meanwhile, a small percentage of men and women appear to have a libido so low as to self-identify as asexual. Again, when a trait is allowed to exist on a continuum, a small percentage of individuals are bound to find themselves at the extreme ends, meaning asexual or hypersexual when the trait in question is sex drive. In that sense, asexuality is bound to exist and to persist. Many asexuals express desire for a romantic, though not sexual, relationship. Most female asexuals identify as female and seem to have a preference for a romantic relationship with a male, and vice versa for male asexuals. Women comprise the majority of asexuals—about two thirds of them. This is not surprising given that women, on average, tend to have lower libido than men. In the absence of reliable contraception—meaning during almost the entirety of human history and prehistory—a lower libido may well have been lifesaving for women given that many routinely died as a direct result of complications during pregnancy and labor.

A final distinction I want to draw is the one between the desire to reproduce, in the sense of wanting to produce children, and the desire to engage in sexual activity for the sake of sexual enjoyment, usually translating into the active avoidance of pregnancy. The effect of the sexual revolution of the 1960s with the advent of highly effective methods of birth control, particularly the kind giving women the ability control their own bodies, has been tectonic in terms of altering our lifestyles—probably on the same scale as the discovery of fire, particularly for women. It has decoupled sexual activity from the commitment to having and raising children. That commitment had been, until then, more or less obligatory and automatic for women, yet biologically optional for men. Social constraints—such as a strict ban of premarital sex enforced through formal laws, religious dictates, or tribal mandates—used to be the main means of enforcing commitment of material resources needed to raise children from men. In addition, men may have evolved an instinctive drive (of varying strength, hence the need for social constraints) to provide such resources. As a result, sexual activity used to be a costly and scarce resource for both sexes, but more prominently for women. Reproductive control has made it both cheaper and more abundant for all. In consequence, there is a growing demographic in the western world consisting of the men and women who like to self-describe as childfree (i.e., not having children by deliberate choice), as opposed to childless (i.e., not having children, despite wanting them, because of biological or environmental constraints). As of 2020, about 10% of men and about 10% of women in the US self-describe as childfree; these percentages are somewhat larger in western Europe. The difference between childfree and childless is choice. The childfree (technically a subset of the childless) do not have children by conscious choice. The childless are those who want to have children but cannot—for reasons as diverse as infertility, lack of a long-term partner, or disability. Whether there is a distinct drive to have children or not is a matter of debate. Yet, based on observation, the desire to have children seems to come on a continuum, just like the rest of our traits. While having children seems to have been a dream and a strong life-purpose since childhood for some, others seem to have felt repulsed by it from an early age, with most people falling in the continuum between these two extremes. In women, anxiety related to the pain and complications of pregnancy and delivery, along with the drudgery of lactation and the round-the-clock care of a helpless infant, can contribute to a desire to avoid pregnancy. The extreme version of this anxiety manifests itself in a condition called *tokophobia*, in which dislike of pregnancy exhibits itself in the form of outright disgust for the condition. Whether early androgen exposure contributes to this feeling is not known.

The overarching aim of this entire chapter has been to highlight the incredible extent of diversity among the individuals who make up humanity. The wonderful feat of modernity has been to recognize this diversity and allow it to flourish, as opposed to denying and oppressing it, as history tells us has been the rule in most preindustrial societies. As diverse as we are, evolution seems to have equipped all of us with specific brain circuitry along with biochemical processes endowing us with the capacity to experience a specific mental state commonly referred to as *being* or *falling in love*. I spent some time describing the individuals who fall in love. I want to devote the next chapter to describing those we are likely to fall in love with, when we are most likely to experience this mental state, and how long we can stay in it once it has been triggered.

3

The Role of Evolution in Determining Who We Fall for, When, and for How Long

Humans have evolved both long-term and short-term mating strategies. Most of us display a stronger preference for one or the other, without being constrained to a single, invariant strategy. Our conscious mating strategy can vary based on our age and personal circumstances. It is also influenced by constraints within our environment over which we may have little control, such as sex ratio imbalances in addition to geographic, macroeconomic, historical, or cultural constraints. The thesis of this book is that, regardless of our primary or preferred mating strategy, falling in love is our evolved mechanism to forge bonding long enough to ensure pregnancy, childbirth, and the rearing of the offspring through at least the most critical years of early life. As such, it is fundamentally a long-term mating strategy whereby a total stranger is rapidly transformed into kin. Moreover, it is a spontaneous process, being almost entirely controlled biochemically. The prefrontal cortex takes a back seat while our brain's love circuitry takes charge, lets the emotions it normally keeps in check run wild, and even assists this circuitry through the generation of fantasies. The process of falling in love is largely unplanned and unconscious. In fact, the choice of the verb "to fall" in the expression "falling in love," which is commonly used to refer to this mental state, highlights the involuntary nature of it. It is generally felt as something that "happens" to us.

As we go about our daily activities—whether at home, at school, or at work—we consciously engage our prefrontal cortex and the areas of our cortex involved in the processing of sensory information. In the background and unbeknownst to us, the love circuitry of our brains—buried deeper under the cortex within some of the brain's most primitive zones—is also active and continuously scanning our surroundings for an optimum genetic candidate with which to produce and rear healthy offspring. When it detects a match at an opportune moment, it triggers the falling-in-love process, and our prefrontal cortex largely yields. As unique as all we all tend to think that we are, we seem to be incredibly similar and predictable in who we are preprogrammed to be most likely to fall for.

In an ideal world where we could have it all, men and women alike would want someone young, tall, physically healthy and strong, mentally healthy and stable, endowed with above-average intelligence and high socioeconomic status, well-connected, powerful, kind, sociable, loyal, tremendously generous, and with an attractive face and body. Most of us, however, fall quite short of this description. And the world we live in is less than ideal. It is, however, the best it has been—at least in its developed parts. For much of history and even longer prehistory, humanity has lived on the brink of starvation, in a world where war, infection, and disease were widespread, with rampant poor hygiene, manual and back-breaking work, commonplace infant and child mortality, regular death in the throes of the mind-tearing pain of parturition, and virtually nonexistent medical help. In such a less-than-ideal world, populated with less-than-ideal people, making concessions in our mate choices was necessary. The inclusion of a delusional aspect in falling in love, recognized even in common parlance through such expressions as "love is blind," may well have been indispensable in simply making reproduction happen. Since the burden of both survival and reproduction—the two driving forces of evolution as theorized by Charles Darwin—was too heavy a load for one, we came to be social animals, relying on reciprocal help within kinship-based tribes of about 150 individuals, comprising a number of nuclear family units, made of a man, a woman, and children. Humans came to be one of the most specialized among all species with respect to these two drives. Women became specialized in shouldering 90% or so of the reproductive toil, spending nearly the entirety of their adult lives going from pregnancy to pregnancy while also sweating and slaving in infant- and childcare in the hope of seeing a few of their offspring survive into adulthood and reproduce themselves. Men, meanwhile, specialized in shouldering 90% or so of the survival toil by supplying the family with essential material resources and protection. This specialization of men and women is reflected in the traditional gender roles we still see today and is technically referred to as "differential parental investment," meaning women would provide biological resources through gestation, delivery, and lactation, in addition to care with regard to the activities of the daily living of the family, while men would provide material resources in the form of food and shelter. The fundamental traditional barter between men and women has thus been that of biological resources for economic resources or—in very simplistic terms—the exchange of sex for food. Our unconscious long-term mate choices reflect the concessions in long-term mate choices that ancestral men and women came to routinely make. These concessions are largely aligned with the notions of differential parental investment, generosity, and paternity certainty. Hence, women will prize above all else generous men who have demonstrated the ability to acquire material resources (as indicated by ascending social status among men) or who actually display large amounts of such resources, while men will prize above all else women who display the best biological resources, namely youth (a proxy for fertility) and physical attractiveness (a proxy for health), and those best showcasing premarital chastity and fidelity. If all this discussion makes it sound like both falling in love and mating are highly transactional at the core, it is because they truly are. When postpartum depression occurs—in women most frequently, but also in men concurrently sometimes—it may, at the root, be caused by the deep realization that the price we find ourselves having to pay to get what we needed far exceeds what we anticipated having to pay. Whether we like it or not, our brains come pre-equipped with the imprints from our ancestral past. Individuals within each sex are acutely aware (albeit mostly unconsciously) of the resources for which individuals in the other sex are primarily looking, and compete among themselves in acquiring and displaying them. Now I want to go into the details of some of the more prominent items in the ancestral wish list our brain's love circuitry is scanning our environment for in the hope of tagging a specific individual as our preferred long-term mate, and then ensuring that we pass our genes into the next generation by triggering our evolved mental state called "falling in love."

Our DNA is the genetic program that codes for every single one of our physical and psychological traits. Robert Plomin refers to it as the blueprint that makes us who we are. More than 99% of our DNA is exactly identical to that of every other human being. Less than 1% of the DNA differences between us is what makes each of us a unique individual. We usually have little trouble accepting the fact that our physical traits are genetically determined to a large extent. But we seem to have more difficulty acknowledging the fact that the same is true of our less-tangible traits. Research indicates that genetics account for 50% of our psychological differences, including our personalities and our mental abilities or disabilities. The remaining 50% appear to be due to largely unsystematic or random environmental factors. We all seem to have an intuitive knowledge of our genetic propensities, both physical and psychological, a knowledge so astute that we unconsciously select, modify, and even create our life experiences to match our genetic selves. What the love circuitry of our brain is really scanning our environment for is our genetic match, the best we can get. When it comes to reproduction, each of us seems to be acutely aware of our mate value in the mating market. We seem to intuitively know that someone of roughly equal mate value is more likely to be willing to mate with us. Thus, 9s generally end up mating with other 9s, 7s with other 7s, etc. When two people in a couple seem to have a large visible difference, there is usually an equally large invisible difference between them to compensate the other way. For instance, I wondered for many years why Catherine Zeta-Jones, one of the most beautiful and powerful stars in Hollywood at the peak of her career, would marry Michael Douglas, another Hollywood titan, one who would have been a handsome match for her had he been some twenty-five years younger, when she could have had any gorgeous younger man her heart desired. It was only when Zeta-Jones came out publicly with her bipolar condition that the pairing seemed to make sense; it was balanced after all.

Individuals in a couple seem to be matched not just on the level of physical attractiveness, but on many psychological traits as well, including personality and intelligence. This is formally called "assortative mating," a form of sexual selection in which individuals with similar phenotypes mate with one another more frequently than would be expected under a random mating pattern. It is well known, for instance, that people with mood disorders tend to form pair-bonds with others who are also endowed with a mercurial temperament. Many theorize that the reason why more and more autism spectrum cases are being identified is the increasing number of marriages between people endowed with a high degree of visuospatial ability and who tend to score low on the ability to emphasize, usually those in STEM fields and those who tend to be endowed with above-average intelligence. We are, thus, most likely to be paired with individuals of similar educational and socioeconomic attainment as us, both of which have strong positive correlations with general intelligence (as measured by IQ or the "intelligence quotient"). Research shows that we do not marry educated people simply because we happen to hang around with educated people; we actively seek them out. One would intuitively guess that assortative mating would make the pair-bond more durable, which is needed given the time necessary to ensure conception, delivery, and the rearing of offspring through at least the first few critical years of life. It is interesting to note here that, in many cases, the idealized image we tend to project on the person we fell for while we were in love is a lot like the opposite-sex version of us. This is perhaps due to our egoistical and innate desire to see in our offspring a near-perfect replica of ourselves so that we may satisfy our unspoken vanity. Or perhaps it stems from a deep but unconscious knowledge that union with someone similar to us is likely to be stronger and last longer. The adage that "opposites attract" is not, however, entirely false. Complementarity can confer benefits by making the couple stronger when considering the sum of its parts, and confer traits to the progeny which may allow for better adaptation to changing environments. Besides, diversity—especially when it comes to the immune system—can mean the difference between survival and death in the fight against pathogens. Today's divorce statistics are a testament, however, to the complexity in striking the right balance in terms of achieving enough similarity for a stronger pair-bond and enough dissimilarity for strategic adaptation to changing environments. In the US, 50% of all first-time marriages and 70% of all second marriages will end up in divorce. Many have naively suggested that arranged marriages may be better than marriages by choice because divorce rates tend to be a lot lower in societies that practice the former. This argument entirely ignores the desperate economic reliance of women on men in such societies and the forbidding price of divorce for their people who tend to have scarce means, leading them to the simple resignation to live in a low-quality union and acceptance that suffering is a necessary part of life.

A distinct, yet related, important and complex task for our love circuitry is to strike the optimum balance between inbreeding and outbreeding in selecting a reproductive partner. Inbreeding is the production of offspring between two individuals who are closely related genetically. Recessive genetic mutations need the presence of two copies to be expressed, one on each chromosome of a pair. Inbreeding increases the probability of genetic disorders in the offspring due the expression of recessive genetic mutations which are harmful. I mentioned in the preceding chapter how each of our psychological traits comes on a continuum and is coded by thousands of point mutations in our DNA. When the extreme version of a trait with strong negative repercussions on quality of life (such as severe mental or physical illness) is present in many members of a family, production of offspring between close relatives will fail to dilute that trait and perpetuate it in its extreme form in the offspring. Note that inbreeding may actually be desirable when the extreme trait running in the family is high intelligence or stunning beauty or above average health.

In addition, a crucial advantage of sexual reproduction over hermaphroditism is immune system diversity to stay one step ahead of coevolving parasites. A particular set of genes called MHC (major histocompatibility complex) is known to be involved in protecting us against pathogens. An experiment called the "sweaty T-shirt experiment" (Wedekind et al., 1995) demonstrated that MHC influences body odors and body odor preferences in humans. Male students were asked to each wear a T-shirt for two consecutive nights. Each female student was then asked to rate the odors of six T-shirts chosen by the researchers. The female students never saw the male students. The women scored male body odor as more pleasant when his MHC was more different from their own than when it was more similar. Humans, just like individuals of most other species, seem to have evolved a natural aversion to incest. MHC as reflected in body odor may be used as a marker for the degree of kinship, so is excessive familiarity with someone. Some amount of outbreeding can, indeed, significantly increase the offspring's fitness. When outbreeding is stretched too far, however, fertility can be adversely affected. The fetus is a foreign organism growing in the womb of its mother. Both the degree of difficulty of the pregnancy and the probability of miscarriage (or spontaneous abortion) are increased when their biochemistry differs too much. A well-known example is Rhesus (Rh) incompatibility between the blood of the mother and that of the fetus. Rh refers to a protein—Rh(D)—found on red blood cells. More than 85% of people are positive for the Rh(D) protein, which is known as being Rh-positive. Some people lack the Rh(D) protein, and they are referred to as Rh-negative. This has no negative impact on their general health. But when an Rh-negative woman is impregnated by an Rh-positive man, there is a high likelihood for the fetus to be Rh-positive. Blood cells from the Rh-positive fetus can enter the mother's bloodstream during pregnancy or more commonly during delivery—where a lot of tearing and bleeding can happen—and trigger the development of Rh(D) protein antibodies in her bloodstream. In the first Rh-positive pregnancy, the fetus is usually not affected as it takes time—about seven days—for the antibodies to develop. In future pregnancies, these antibodies can pass through the placenta and attack Rh-positive red blood cells in the developing fetus. This can lead to miscarriage or hemolytic disease of the newborn who, as a result, can fail to survive. It is now routine practice to give Rh-negative mothers an injection of Rh immunoglobulin before and after childbirth to prevent her from developing Rh(D) antibodies. This example serves to illustrate how biochemical similarity between mother and baby may ensure a homeostasis needed for a comfortable and completed gestation. Our body synthesizes many biochemicals with powerful effects on our well-being. One can intuitively guess how a mother's functioning could be affected by the biochemistry of the fetus and vice versa. As common as interracial marriages have become today, we are still most likely to fall in love with unfamiliar individuals from our own ethnic group. The latter seem to be the most likely candidates in terms of achieving optimum compromise between inbreeding and outbreeding.

Since they relied on men for both their survival and that of their offspring, women came to value generosity in men as an eminently important trait, perhaps more important even than displays of actual resources or ability to acquire them. An ancestral woman would have found herself in the position of a lifelong beggar with a miserly and callous husband. Even today, when a man takes a girlfriend on an expensive dinner for which he insists on shouldering the entire cost or when he buys her an expensive engagement ring, he is displaying his propensity for generosity to the woman who, he hopes, will agree to bear his children. Many preliterate communities still enforce the custom of "bride price." This price oftentimes includes expensive jewelry for the bride-to-be, besides money or other economic resources given to her family. Men of scarce means will often give away what amounts to the entirety of their wages for many months of labor. Perhaps the most important goal of paying a bride price is to demonstrate generosity.

Since they relied on women to carry, deliver, and care for their children, men came especially to value displays of fertility and health in women. Perhaps more prominently though, they came to place an obsessional importance on ensuring paternity certainty. Ancestral men had little appetite for toiling their lives away for the sake of raising another man's children. In contrast, maternity has always been certain. The advent of reliable birth control, anonymous urban living, and financial independence has largely liberated modern women from the tyranny of traditional female-female competition which, in one of its forms, consists of providing behavioral guarantees of paternity certainty, typically with displays of premarital chastity and strict marital fidelity. Where extant, this specific form of competition between women only provides fuel to the jealousy and violence natural to men and further aggravates their own condition. For ancestral men, premarital chastity was the best predictor of marital fidelity. Sacrificing the availability of casual sex was a small price to pay for what they stood to gain by enforcing it. In addition, Roy Baumeister and Jean Twenge have suggested that women may have derived a direct economic benefit for themselves by forbidding premarital sex for other women in a cartel-style price-fixing scheme whereby they were able to extract a higher price for sex from men by making it a rare commodity (Baumeister and Twenge, 2002). Otherwise stated, women have worked to stifle each other's sexuality because sex was a limited resource that women used in order to negotiate with men, and scarcity gave women an advantage. Note that, in ancestral societies where women could not provide for themselves and no public welfare system was in place, this argument would also have provided a strong reason for male family members of those women—meaning their fathers and brothers—to suppress their sexuality, since they would have been those ultimately left on the hook for "feeding the old maid" for life should she be left unmarried. Worse yet, if married to an absent or incompetent provider, they would also have had to feed her children. As both men and women stood to substantially gain from it, suppression of female sexuality came to be the universally observed phenomenon it is, even today. Add to this each woman's self-interest in protecting herself, and altruistically, other female friends and family from the ills of unplanned pregnancy and we get a perfect illustration of the "all roads lead to Rome" adage.

Within marriage, men would typically resort to sequestration and direct violence in order to control their wives' sexuality. Prior to marriage, the use of indirect violence in the forms of slut-shaming and gossiping—formally called "information war"—has been a universal form of female-female competition. Direct violence usually came from close family members in severe cases of premarital infraction. The goal of sexually tainted gossip was, sometimes, to inflict defamation severe enough to result in the banishment of the targeted individual from the marriage pool for life. Note that this also resulted in severe economic harm to an ancestral woman, who was typically dependent on men for her survival. The tactic was eminently effective when life was restricted to the confines of a small village. As a result, this form of competition tends to be both more intense and more ostentatious in preliterate communities, where men, too, are active participants in the community-wide monitoring of fertile women, slut-shaming, and gossiping. The adage that accomplishing certain tasks "takes a whole village" quite literally applies here. It is interesting to note, for instance, the sheer number of female first names derived from Greek with the meaning of "sacred, chaste, pure, maiden, or virgin"—including Aggie, Agnes, Agnetha, Agnieszka, Annis, Aune, Caitlin, Catriona, Cajsa, Carin, Catalina, Catia, Catrina, Cora, Coralee, Coralie, Coreen, Corella, Coretta, Corinne, Corissa, Corlene, Ines, Jagienka, Jagna, Jagusia, Jakayla, Jekaterina, Kaarina, Kaia, Kaisa, Kaj, Kakalina, Kalina, Kaley, Karen, Kari, Kasia, Kata, Katalin, Katariina, Katarzyna, Kate, Katelyn, Katerina, Katherine, Kathleen, Kathy, Kati, Katka, Kato, Katri, Katrina, Katya, Kay, Kayla, Kaylee, Kitti, Kitty, Pallas, Parthenia, Parthenope, Senga, Trijntje, Trina, Trine, and Yekaterina. I am not going to list those derived from Latin with the same meaning; I think you get the idea. Customs extant in certain parts of the world are testimony to the deplorable consequences of the practices mentioned earlier. In parts of the Middle East—certain areas of Afghanistan come to mind among other places—women and girls are almost entirely cut off from the outside world, sequestered inside houses, left to linger in complete ignorance, venturing out rarely, and that, only if covered from head to toe in oppressive burkas and accompanied by a male family member. They spend their entire lives in what amounts to an open-air, maximum-security prison. If it is determined that a new bride is not a virgin, she is immediately shamed and returned to her parents, assuming she is still alive. Virginity is ascertained by the presence of blood stains on the nuptial bed sheets—which is, medically speaking, an extremely dubious method since the hymen can tear for many reasons or be partially present or even absent from birth. It is interesting to note that the mother-in-law is typically the leader of the troop making the assessment, supporting the argument of Baumeister and Twenge about a female cartel out there to enforce a price-fixing scheme. Male martyrs are promised forty virgins to enjoy in paradise—which shows the male interest in ensuring premarital chastity and also begs the question of whether those virgins are also there to enjoy paradise or simply there to live their version of hell. The modern notion of consent, obviously, does not exist under these circumstances. The same communities also provide a glimpse into the extreme forms that differential parental investment and strict division of labor can take. Their women are helplessly dependent on men for their survival. Their men barely speak to their children or wives, let alone to any other woman. Note that all of this represents one step up in civilization compared to societies that still resort to female genital mutilation as a way of controlling female sexuality. Different though it may be, where we come from is not a pretty place! Let's take a moment here to thank science for making reliable contraceptives a reality and for transforming our lifestyles for the better.

Philosophers have long argued that there seems to be something universal in our sense of beauty. Darwin made the point that animals seem to display an appreciation for beauty which is quite similar to that observed in humans. Modern science is proving that beauty is not just in the eye of the beholder. When multiple individuals are asked to rate the same set of faces for attractiveness, clear agreement emerges among individuals both within a given culture and cross-culturally. Beauty is really a proxy for health and an ancestrally trustworthy indicator of genetic quality in humans as in other species—although the enormous progress made in modern cosmetic surgery makes the trustworthy aspect of it more and more questionable. The global market for cosmetic surgery is projected to exceed $22 billion by 2023 and is almost exclusively catering to women. As for the cosmetics and beauty products market, it generates revenues already exceeding $85 billion in the US alone! Its consumers are, once again, almost exclusively female. Both industries have increasingly higher numbers of male customers, which is a great benefit to women, since they get to enjoy the company of better-groomed men. Most of the alterations which women wish to achieve are focused on their faces. This points to the fact that displaying a beautiful-looking face has ancestrally been a prime focus of female-female competition, hence a prime component of male preference. This fact is very much imprinted into the modern female brain. You may dislike cosmetic surgery but, used in moderation, it can significantly improve women's self-esteem and life satisfaction. Speaking about the topic of teenage girls, Louann Brizendine points to their newfound form of self-expression in the form of attracting male attention as the hormones of puberty surge. They will spend hours in front of a mirror inspecting pores, plucking eyebrows, or applying makeup. She adds that girls would probably be doing some version of this whether the media were there to influence their self-image or not. It is a predisposition which is most likely etched into their brains prenatally and gets activated with the hormones of puberty. These teenage girls are no fools. Beauty is strongly correlated with upward economic mobility, especially for women. In ancestral times, it may well have been the only way for women to achieve wealth short of being born into it. Beauty is strongly liable to the halo effect, whereby a beautiful person is automatically judged to be more intelligent, more competent, more resilient, more sociable, healthier, and so on, than a less attractive person. These benefits also accrue to attractive men. And let's be clear: the female brain scans for beauty, just like the male brain does. Ancestral women often sacrificed it, however, for the benefit of having a generous long-term partner endowed with material resources or a high social status. One cannot have it all.

What is interesting is that science shows a linear relationship between any two of the four variables consisting of beauty, IQ, physical health, and mental health within the normal range, meaning between plus/minus one standard deviation of the observed population mean for each variable, validating some of the stereotype. This is technically referred to as "overall system integrity." Where the stereotype gets it wrong is at the low end and high end of the range for these variables. The overall relationship is S-shaped, meaning people judged least beautiful tend to have higher intelligence and better physical and mental health than what one would naively expect. Those judged most beautiful are less intelligent and less healthy, both physically and mentally, than typically assumed.

One indicator of beauty our brains have been conditioned to scan for is left-right facial symmetry (Little et al., 2011). Facial asymmetry is a particularly useful measure of developmental stability because we know that the optimal developmental outcome is symmetry. The presence of asymmetries can point to a number of developmental issues including malnutrition, the presence of a high parasite load, deleterious mutations, and inbreeding. It is a direct indication of the genetic quality of our parents, hence an indirect measure of ours since we inherited a combination of their genes. Malnutrition in the womb can be indicative of parental incompetence with respect to the acquisition of material resources and low social status. The presence of infections points to a defective immune system or poor hygiene, which can indicate anything from low social status to mental illness, intellectual or physical disability, laziness, or simple ignorance. Excessive amounts of deleterious mutations and inbreeding could point to a severely limited pool of reproductive partners for both our parents, hence to their place at the bottom of the mating totem pole. Preference for symmetry can then potentially confer direct benefits to the astute chooser, such as avoidance of contagion, as well as indirect benefits in the form of quality genes passed onto offspring.

A second facial feature that our love circuitry is looking for is averageness, which refers to how closely a face resembles the majority of other faces within a population. An average male face can be generated by a computer by making a composite of a large number of male faces within a big city. Such a face is usually judged attractive by observers. Averageness is a sign of underlying genetic diversity. Remember, genetic diversity helps to stay ahead of the arms-race against parasites. Facial averageness has also been found to have a positive correlation with medical health from actual medical records in both men and women. Who would not prefer to be partnered with a disease-free individual, especially if it is for the long run?

A third facet of beauty we value in a long-term mate seems to be the display of extremes of secondary sexual characteristics (more feminine for women; more masculine for men). Faces of men and women are similar in childhood and begin to diverge starting with puberty under the influence of sex hormones as secondary sexual characteristics are bestowed. A more feminine-looking face (with both plump limps and plump cheeks, resulting in a more round face) may be preferred because it advertises high levels of estrogen, a sign of fertility. A more masculine-looking face (with larger jawbones, more prominent cheekbones, and thinner cheeks), in contrast, advertises a strong immune system and health. This is because testosterone is known to suppress the immune system, so the cost of growing the traits means that only healthy individuals can afford to produce them.

A fourth indicator of facial attractiveness consists of the skin's color and texture. Smoothness of skin texture and homogeneity of skin color are found to be positively correlated with ratings of attractiveness. Both are indicators of health (in addition to being indicators of age). Indeed, many illnesses and infections can result in rough, patchy-looking skin. In short, a woman's facial beauty contains cues about her health and, in some respects, her fertility. As a result, men came to prize it highly when looking for a reproductive partner. After all, successful reproduction requires someone who cannot just become pregnant and produce healthy offspring but, just as importantly, someone who can sustain the damage (both physical and mental) from the assault of multiple pregnancies and deliveries. This may actually be the main evolutionary selection criterion to account for the universal fact that women live longer lives than men. They seem to be able to withstand the assault from the multiple diseases of aging for many more years than men can.

Another major item that the male brain's love circuitry is wired to look for in a long-term mate is youth, which it uses as a proxy for fertility. Fertility decreases with age for both men and women, but more dramatically for women. Men's fertility peaks in their mid-to-late teens and slowly decreases into old age. A sixty-year-old man may not have the highest quality sperm, but he can still impregnate a younger woman (assuming that he is one of the roughly 40% of men in that decade of life not affected by erectile dysfunction, although the blue pill can come to help today). A woman's fertility rises starting with her mid-teens, reaches its peak around her mid-twenties, slowly decreases into her mid-thirties, then plummets until it is essentially nil by her mid-forties. This happens despite the fact that many women still menstruate into their fifties; the average age of menopause in the US is roughly fifty-two. This has made youthful appearance an ancestrally reliable indicator of fertility, especially for women. I mentioned the fact that a round, more feminine-looking face hints at fertility. In addition, the unaltered texture and color of facial skin prominently hint at age, so do the natural texture and color of hair. Again, modern cosmetic surgery in conjunction with state-of-the-art cosmetic products are starting to make this reliability more and more questionable. This is actually a major improvement in the life quality of the older women who can afford them. Women who wish to continue being sexually active and adventurous into their sixties or seventies now have more leeway in doing so. Many of Hollywood's actresses in their fifties today could compete with any twenty-something woman for looks. The amount of frustration that some older women must have felt in ancestral times is unfathomable. As recently as the early nineteenth century a thirty-year-old woman would have been considered to be at the threshold of old age!

Fertility treatments are also available to help couples conceive. This, again, is great news for today's men and women who wish to have children later in life. Ancestral times have, however, left their imprint on our brains. Cross-culturally, husbands are on average three years older than their wives. Men seem to prefer younger women as marriage partners, but not too young. This may be because they want to strike the right balance between fertility (signaled in youth) and similarity in terms of maturity and life stage to ensure a stronger pair-bond and marriage stability. Women seem to prefer older men as husbands, but not too old. They may do so in order to strike the right balance between giving men sufficient time to develop the skills needed for status or resource acquisition and achieving similarity needed to form a longer-lasting pair-bond.

Having an hourglass body shape provides another major cue to a woman's fertility. The best metric to assess it is called the "waist-to-hip ratio." Indeed, research shows that body fat distribution as measured by waist-to-hip ratio is correlated with youthfulness, reproductive hormonal status, and long-

term health risk in women (Singh, 1993). The ideal waist-to-hip ratio for
women is determined to be roughly 0.7 cross-culturally. Fat distribution
in humans depends both on sex and age—the sexes are similar in infancy,
early childhood, and old age, and differences in fat distribution are greatest
from the early teens until late middle-age. The biggest differences in the
physiology of fat accumulation and utilization are observed in the abdominal
and buttock-thigh regions. Testosterone stimulates fat deposits in the former
and inhibits fat deposits in the latter region. Estrogen, in contrast, inhibits fat
deposits in the abdominal region and stimulates fat deposits in the buttock-
thigh region more than in any other region of the body. Married women
with higher waist-to-hip ratio and lower body mass index report having
more difficulty becoming pregnant and have their first live birth at a later
age than married women with lower waist-to-hip ratio. Low waist-to-hip
ratio also accurately signals health as defined by absence of major diseases.
A large number of studies have found that the risk-factor profile for major
obesity-related diseases such as diabetes, heart attack, and stroke varies
with the distribution of fat rather than total amount of fat. Besides body
fat distribution, a larger pelvis in women relative to men contributes to
their hourglass shape. A larger pelvis in women signals relative ease during
delivery and, ancestrally, a smaller likelihood of death during childbirth. It is
not surprising that liposuction has come to be the popular procedure it is
today. Breast augmentation in conjunction with abdominal liposuction can
enhance the hourglass look of a woman and make her look more attractive
to potential mates. Note that bigger breasts really mean plentiful nutrition
for the future baby. It probably hurts men's feelings a bit to think that this is
the reason behind their obsession with larger breasts in women, but it truly
is.

Natural aging robs us of beauty, fertility, and health. All of us can easily
distinguish an older person's voice from that of a younger person without
seeing them. We may have evolved this ability in eras that predate the control
of fire where humans were left in total darkness soon after sunset. Research
even shows that people's ratings of attractiveness based on voice alone
correlates positively with their ratings of facial attractiveness. This is not

something cosmetic surgery can currently correct for. Skin tags and moles grow bigger and multiply with age. Most people find wrinkles repulsive. They make people look eternally sad, which is not attractive psychologically. Older people's faces and bodies also become less symmetric. Their movement and thinking slow down. Some shrinkage in stature also occurs as we age. The above is only a short list of the ways that our age is being signaled to others. Nature seems to be doing its very best to prominently advertise age in every way possible. The message, from a reproductive perspective, is quite clearly not to waste any time or precious resources on compromised genetic quality and low fertility.

Women's brains, meanwhile, are strongly wired to look for signs of resource acquisition abilities or the outright display of such resources in a long-term male partner. Women prize the notion of the ability to acquire resources somewhat more than current resources in possession. Ability ensures stability of provisioning over an extended period of time. Current resources, unless vast or periodically replenished, may become exhausted over time. Three indicators have prominently been used by ancestral women to gauge men's ability to acquire resources: their status among other men, their V-shaped torso, and their height. In the absence of these indicators, but more likely in conjunction with them, men could count on the ostentatious displays of actual material resources or wealth in making themselves attractive to fertile women. Indeed, a woman can increase her reproductive success by choosing a high-status man who controls resources and, hence, can provide material security to successfully raise her offspring (Singh, 1993). Of all indicators, women evolved to prize a man's social status most. Men achieve status by competing with other men. According to David Buss, men (but not women) had assigned ranks even among some of the most primitive tribes known. Men with the highest ranks were given the most fertile and beautiful women to impregnate. Deep in the human evolutionary past, men may have used direct physical confrontation to achieve status. The fact that they possess, on average, a larger body (thicker bones, higher muscle mass, taller frame, deeper voice) than women is testimony to the fact that these characteristics have been preferred by our most ancient foremothers. These characteristics were later co-opted to the acquisition of resources by brute force. As intelligence came to outweigh brute force in resource acquisition, women may have started to evolve a preference for men whose intelligence was in the average to above-average range. In hunter-gatherer societies, the better hunters may have been those with higher intelligence. In addition, being endowed with a larger torso may have conferred a physical advantage in hunting, hence the evolved female preference for a V-shaped body in males, meaning broader shoulders in relation to their waist. The waist-to-hip ratio can also be used in the quantification of what is meant by a V-shaped body. Cross-culturally, the ideal waist-to-hip ratio for men appears to be roughly 0.9. When it comes to height, men are on average six inches taller than women in the US, where the mean male height is about 5'10" and the mean female height is about 5'4". This results in a metric called "sexual dimorphism in stature" (defined as: SDS = male height/female height) of about 1.09. A research study (Pawlowski, 2003) shows that people adjust their preferences for SDS in relation to their own height in order to increase the potential pool of partners. On online dating sites, profiles of men who are of average height or taller are viewed more often than those of shorter men. Taller men get more dates, pair up with more attractive women, and even score more promotions and higher pay at work. Shorter men tend to marry younger and less educated women. Since tallness is associated with higher social status and dominance, psychologists think that shorter men may unconsciously be trying to compensate. They also tend to marry later and to divorce less frequently than taller men.

With the onset of puberty, when teenage girls become increasingly concerned with their looks and obsessed with participating in sexually tainted gossip, teenage boys become more distressed by challenges to their authority. On average, boys also start to spend more time at the gym to achieve the ideal shoulder-to-hip ratio and convey strength. They wish they owned Ferraris or Lamborghinis to impress women by displaying wealth. These tendencies remain stable with age. Indeed, while many middle-aged men choose to spend their discretionary dollars on luxury cars, same-age women typically prefer to buy a bundle of cosmetic procedures instead. This is very much in line with the concept of differential parental investment, traditional gender roles, and the ancestral barter between the sexes of sex for food. The vexing part for us today is that knowing this does nothing to change these deeply rooted preferences. We just keep gravitating toward them, even consciously. They are "the achievements"—because they inspire us with the feeling of having bested our reproductive competitors—for which our brains are preprogrammed to reward us with the feelings of pleasure and contentment. Of course, with even more discretionary dollars, men and women alike may opt for both sets of luxuries.

Just as we have inherited our intuitive preferences in reproductive partners from the humans who preceded us, so have our brains been bequeathed with the tendency to make us fall in love when it has been most opportune for those who came before us. Contrary to the adage, love does not strike at any time. It only feels that way because it happens unexpectedly. We have to remain alive in order to reproduce but, once we have done so, we need to remain alive to help our offspring stay alive long enough to reproduce. So the best moment for our brain's love circuitry to make us fall in love would be at the confluence of the two fundamental evolutionary drives—namely survival and reproduction. We tend to reproduce when doing so either helps us to survive, or when it hurts our efforts to survive the least. The former tends to be the case for women more often than not, and the latter is more often the case for men. It should go without saying that the converse may also happen, albeit less frequently. The moment when we are most likely to fall in love is pertinently illustrated by the damsel-in-distress archetype found in many folk stories, children's books, novels, and movies. In this archetype, the damsel falls in love when her survival is seriously threatened and her only chance at salvation is to accept her suitor's invitation for sex—ahem, for marriage, following good etiquette. Notice that this pretty much implied accepting to bear his child in the absence of reliable contraceptives, meaning for almost the entirety of human existence. The suitor is, of course, a prince or someone endowed with wealth so vast that supporting a wife and all the children she may bear is nothing but a trifle.

Think about the storyline for Cinderella, a children's tale most of us grew up with. It not only gives the gist of the ancestral wish list for both men and women with regard to the qualities we tend to seek in a long-term mate, it also illustrates very simply and clearly *when* falling in love happens. Even children recognize the truth, which may account for the story's worldwide appeal and popularity. Cinderella is young, beautiful, kind, and humble. Her youth stands for fertility, beauty for health, kindness for the propensity to care or attend to the activities of the daily living of vulnerable others (meaning children, the elderly, and the invalid in the family), and humbleness for chastity and fidelity. Her humility ascertains her ability to withstand the rigors of lifelong sexual frustration, the traditional fare of "being a good mother." I could have substituted the word "virtuous" for "humble" in describing her. The propensity for chastity and fidelity is what was typically meant by the word "virtuous" when used in relation to a woman. Humility is also associated with the willingness to submit which, again, was something expected of a wife. Cinderella is of noble descent, meaning she comes from good genes, and matches, at least in birth, the prince's social circle. They are, however, unfamiliar to each other. This means that they may be well matched in terms of striking the right balance between both similarity and dissimilarity and between inbreeding and outbreeding. She fell into abject poverty after her father and original protector passed away, leaving her at the mercy of her cruel stepmother and equally cruel stepsisters. She is employed by them in menial domestic duties, the skills needed in motherhood, therefore traditionally prized in young girls of marriageable age. She has everything the prince—or any of his male contemporaries—could wish for in a future wife. The prince, on the other hand, has the highest possible social status besides the king. In fact, he is destined to become king himself very soon. He is young, handsome, courageous, strong, and immensely wealthy. The characteristics that are, however, the most highlighted are his social status and wealth. He is willing to commit his vast resources to a single woman and her offspring. Indeed, he has publicly announced his intention to marry. Who could best him for being generous? He has everything Cinderella—or any of her female contemporaries—could wish in a husband. They meet when she is in distress. She is impoverished, abused, defeated, and desperately in need of help. He is ascending in social status, as he is soon to become king. He turns into her savior. They fall in love at first sight while dancing together at a ball he set up to choose his bride-to-be, marry, and live happily ever after. Behind Cinderella's socioeconomic helplessness lurks the idea of sexual exploitability. Behind the prince's consent to commit his vast resources lurks the idea of economic exploitability.

Both protagonists are in an intense emotional state. Their respective emotional state is, however, specific to their gender. Cinderella is riding an emotional low. She is under tremendous economic duress. She is probably feeling very saddened by her current situation, extremely anxious about her future, and desperate for help. In addition, she has been in these dire circumstances for an extended time period, probably several years. Her worn-out brain's love circuitry is ready to fire. One survival strategy that the female brain has evolved is the exchange of sex for material resources. After all, sex has come at a pretty high biological cost to women, one that has seriously hampered their efforts to fight for their own survival. Indeed, they have had to endure the crippling effect of pregnancy (with no way to control it), the pains of childbirth (and manage to survive it without medical assistance), and the drudgery of lactation and of all the menial tasks associated with caring for babies, infants, and children. They were bound to exact an exchange. Therefore, they would have been most likely to consent in times of socioeconomic duress, when their drive to survive meshed best with their drive to reproduce. The prince, on the other hand, is riding an emotional high. He has been waiting to become king for many years. He has had to go through a lengthy education and grueling training. He has had to prove himself to his father, maybe by enduring war. The king is finally ready to cede the throne and wants to see his son married. The prince will be catapulted to the highest possible social rank. He is feeling overcome by joy, pride, and success. He has been riding this emotional high for an extended time period, probably a few years. His overwhelmed brain's love circuitry is ready to fire. One reproductive strategy that the male brain has evolved is the exchange of economic resources for sex. Men would have been best positioned in that bargain when they had achieved the social rank to command control of enough resources needed to support a family, meaning in times of socioeconomic success, which would have been when their drive to reproduce meshed best with their drive to survive. It follows that women are most likely to fall in love when they have been in an intense emotional low for an extended time period—an emotional low typically caused by seriously jeopardized economic prospects. Men are most likely to fall in love when in an equally intense emotional high for an extended period of time—an emotional high usually elicited by considerably improved economic prospects. The length of the time needed to get our brain's love circuitry to fire probably varies with individuals. Those whose neurochemistry is more likely to go out of kilter with ease probably fall in love more quickly and more frequently during their lifetime. Those whose brain chemistry is more rigidly controlled may need the duration of the heightened emotional state to be longer and its intensity higher before their brain triggers the falling-in-love process. As a result, they may fall in love once, maybe twice, or maybe even never during their lifetimes. Note that the latter case would have no bearing on whether they would eventually marry and/or have children or not. Falling in love is not necessary for any of that to happen.

How long we stay in love is intimately connected with the length of time needed for human brain development. As our brains became bigger and ever more complex throughout our evolution, human babies came to be delivered sooner and sooner, hence in an increasingly vulnerable and dependent state. Childhood also lengthened, making it necessary to tie parents together for longer periods of time. As a result, it is highly likely that we came to remain in love for an increasingly longer time stretch. A walk down the anthropology lane can help us understand how the human brain's love circuitry may have evolved into its modern form.

Long-term sexual relationships in humans have two components: the attraction phase to compel a male and a female into having sex so that conception can happen, and the much longer attachment phase which drives them to remain together in order to raise one or more children into self-reliance. The attraction phase, sometimes called "passionate love," probably emanates from some of the most ancient circuitry of our brain. The first discovery of hominid ancestors living in trees was made in Chad, central Africa, in 2002 (Brunet et al., 2002). The associated fauna suggested that the fossils were between six and seven million years old. It was not possible to determine whether they were bipedal or walked on their four limbs. These ancestors probably possessed some of the same brain circuitry for attraction that we still possess today. Maybe around that time, maybe somewhat later, hominids became bipedal. Walking freed their hands to hurl rocks at small animals, allowing them to kill more efficiently and eat more meat. Note that it also enabled them to hurl rocks at each other and kill off their reproductive competitors. Helen Fisher suggests that female hominids may have first started to need males for food and protection around the time that they became bipedal. Indeed, they now had to carry their babies in their arms instead of on their backs, which was a handicap when they had to gather food or self-protect. She also suggests that this may have been the very beginning, for humanity, of the notion of fatherhood. Most animals do not recognize their fathers; they only recognize their mothers. So our mechanism of attachment to a sexual partner, also called "companionate love," may have started to develop around that time. Research shows that the attachment system that originally evolved to forge bonding between mother and baby may have been co-opted to the process of bonding with our sexual partners. It is therefore highly likely that females evolved it first and passed it down to males over time. This would also explain why modern females tend to attach faster and more strongly than males. Analysis of fossilized remains from the *Australopithecus afarensis* dating about 3.5 million years ago leads to the conclusion that they were principally monogamous, since they displayed about the same amount of sexual dimorphism as today's men and women. These forebears probably already had an attachment system resembling ours. They were bipedal. Their cranial capacity was only slightly larger than the average cranial capacity of today's chimpanzees. They could not yet speak. Archeologists have not found any cave art or tools they might have used. For males, gaining an edge in courtship probably required advancing in status among other males, as is the case for the male in many other species of mammals. Human prehistory truly started with *Homo habilis* (meaning the "handy man") some two million years ago. These human forebears crafted stone tools and knives which they could use to hunt for big game. They assembled at meat processing areas where they would cut the large animals they had killed and share it among themselves. Consuming large amounts of meat is probably what powered the tremendous growth of the human brain thereafter. Indeed, building and maintaining a large and complex brain is a process that puts enormous caloric demands on a biological system, far surpassing what is normally required to build and maintain internal organs such as the heart, lungs, and liver. As a result, animals with a vegetarian-only diet cannot do both. Human evolution gathered steam as a direct result of consuming more meat. Broca's area, our brain's speech production zone, seems to have started to take a human shape around 1.8 million years ago, from which one may assume that our *Homo habilis* ancestors spoke some form of primitive human language. The courting process probably changed as a result of this newly acquired ability. The first remains of our forebear called *Homo erectus* (meaning the "erect man") were discovered by Kamoya Kimeu, a fossil collector, in 1984 on the bank of the Nariokotome River near Lake Turkana in Kenya (Brown et al., 1985). They were dated to be 1.5 to 1.6 million years old and belonged to an eight- to twelve-year-old boy who was called "Turkana Boy" or sometimes "Nariokotome Boy." Judging from his nearly complete skeleton, his hands, arms, hips, and legs were quite similar to ours. He was tall and probably dark-skinned. If not for his face, he may have passed for a modern human. His forehead was shallow and sloped, his eyebrow ridges thick, his face protruded, his teeth were big, and his chin barely extant. Our *Homo erectus* forebears made more sophisticated tools including hand axes, picks, and knives. They hunted for even larger game such as buffalo and wild boar. Such hunts could only be done successfully by large groups of well-coordinated men with improved spatial navigation skills. They had also mastered fire. Together, they could cut their game, cook the meat, and share a meal. The control of fire had revolutionized their lifestyle. They no longer needed to go to their shelter and sleep shortly after sunset. The ability to socialize around a bonfire well into the night quite possibly resulted in tremendously improved social and language skills.

Another eminently important benefit which came with the control of fire was the ability to cook food, especially meat, which made it tastier, easier to digest, and safer to consume. This meant that our *Homo erectus* forebears could eat more of it. As a result, their cranial capacity doubled compared to that of *Australopithecus afarensis*, attaining in some cases one thousand cubic centimeters—right at the low edge of the one thousand to fifteen hundred cubic centimeters cranial capacity range for modern humans. This is probably when the newest addition to the human brain, the prefrontal cortex, was tied into our brain's ancient attraction circuitry and tasked with assisting the latter by generating the fantasies which serve to amplify the intensity of our emotions when in love. This added element is most likely the biggest difference between the way that we experience this particular stage of the mental state compared to the lower primates. A bigger brain was a boon for humanity, but it posed a problem for women known as "the obstetrical dilemma," which is thought to have started with the moment that the cranial capacity of *Homo erectus* reached around eight hundred cubic centimeters. This hypothesis states that, in order for women to remain bipedal, their pelvic size could not grow any wider. Yet, they had to be able to deliver their big-brained babies through their narrow birth canals. So they started to deliver their babies sooner and sooner. Assuming that this hypothesis is true, one can only imagine what it took for natural selection to do its work and cull those among them who couldn't deliver their baby at an earlier stage of development. A competing explanation for the earlier delivery of human babies is the "metabolic crossover hypothesis," which proposes that energetic constraints of both mother and fetus are the primary determinants of gestation length and fetal growth in humans and across mammals and that there is little evidence that pelvic constraints have altered the timing of birth (Dunsworth et al., 2012). By nine months or so, the metabolic demands of a human fetus threaten to exceed the mother's ability to meet both the baby's energy requirements and her own, so she delivers the baby. Note that the size of our brains directly influences those caloric demands, since our big brain is a major energy-consuming organ. By some estimates, a modern human fetus would have to undergo a gestation period of eighteen to twenty-one months, instead of the usual nine months, to be born at a neurological and cognitive development stage comparable to that of a chimpanzee newborn (meaning at a cranial capacity that is about 40% of its adult size instead of 30% as observed in human babies). This suggests that a modern human baby is born from nine months to a year too soon, which is the reason why human newborns are particularly helpless and vulnerable. During their first year, infants spend 50% of their metabolic energy to nourish and develop their brains, which leaves the latter vulnerable to anything interfering with its delicate bioengineering. The female *Homo erectus* was now more crippled than ever. Indeed, she now had to supervise her baby 24/7, which made her even more dependent on her male partner for her survival and that of her offspring. To make matters even worse for women, childhood started to become lengthier around that time. This is called "delayed maturation." Modern humans need about eighteen years for full completion of puberty. Chimpanzees only need ten years to reach the same stage. Parenthood thus became a two-person job.

Those women and men who could develop a stronger biochemical attachment to enable a longer-lasting pair-bond saw more of their children survive into adulthood and reproduce themselves. This meant that women had to become even more careful in their mate selection. They may have evolved superior theory of mind as a result so that they could better assess the personality and intentions of a suitor. This is also when delayed activation of the love circuitry in women compared to men may have evolved. Delayed activation meant a longer courtship and more time for women to make their assessment. They also had to become increasingly more sensitive to signs of abandonment in their intimate relationship, lest they and their offspring perish as a result of it. This last point is central to understanding why such a large fraction of modern women (up to 30% of them by some estimates) suffer from anxiety or depression. Most of the anxiety they report experiencing is in the form of separation anxiety, which is directly tied to this fear of abandonment, of which they seem to have inherited an extreme version from our foremothers. Scientists have speculated that women may have evolved depression to serve a social motivation function: the high personal costs associated with the anhedonia and psychomotor perturbation of depression may have been useful in persuading their reluctant social partners to provide them with help or to make concessions. This refers to both their reproductive partner and other members of their tribe, and was accomplished via two possible mechanisms—namely honest signaling of distress, as well as passive, unintentional fitness extortion (Watson and Andrews, 2002). In a way, the same idea can be applied to any severe episode of mental illness, especially to those within the psychosis spectrum. Regardless of its nature, mental illness can be viewed as a cry for help from an overwhelmed brain. If we extend this idea to the damsel-in-distress archetype we discussed earlier and take into account the psychotic nature of being in love, we can view this mental state as a Mayday signal for survival from a distressed female brain or an outreach for reproductive help from an emotionally high-strung male brain. The divergence of the lineage leading to *Homo sapiens* (meaning the "wise men") out of ancestral *Homo erectus* is estimated to have occurred in Africa roughly 500,000 years ago. The earliest fossil evidence of early *Homo sapiens* was discovered in Africa and is dated to be around 300,000 years old. These sage ancestors of ours had cranial a capacity comparable to that of modern humans. The love circuitry they came to possess by that time was probably very similar to what we have today.

Love's ability to bring together and keep together human beings has played a crucial role in the survival of our species. Our brain's love circuitry and its associated neurochemistry have been wired into our biology by the forces of evolution to promote the reproductive success of our ancestors. I have examined the environmental cues and circumstances likely to trigger the complex neurobiological phenomenon we call "falling in love" and discussed how natural selection has co-opted the attachment system originally built to sustain the bond between mothers and their infants to forging and maintaining the human adult pair-bond. I would be remiss if I did not give a psychological account of the intensely emotional topic of love. In the next chapter, I will present a psychodynamic model of how falling in love changes our feelings, perception, thoughts, and behavior over time to accomplish its ultimate goal of perpetuating the human species.

4

The Psychodynamic Model of Falling in Love

The psychodynamic model I am proposing for the mental state we call "falling in love" or "being in love" has four stages:

1. A split-second rise in emotional intensity when our brain's love circuitry first starts to flood our system with the biochemical cocktail of attraction.

2. A linear emotional amplification fueled by fantasy as we interact with the loved one during courtship and, if the latter is successful, as we repeatedly have sexual intercourse with him or her. This is the core stage of falling in love, a stage where we slowly enter a state of mild psychosis which can vary in length from two to three months to two to three years in duration—the length of time being determined by the average time it took men and women ancestrally to pair up and achieve pregnancy.

3. A rapid decline in emotional intensity, again within a split second, down to a small percentage of the peak value, maybe a mere 10% or so of the highest intensity reached. What triggers this stage can be rejection, achieving pregnancy, or a simple time-out in the absence of both.

4. A slow exponential decay of the remaining emotional energy taking

place over four to five years with the goal of keeping the pair-bond intact long enough to get us through the pregnancy and the first few critical years of the newborn's life.

And then we are sort of left on our own. What remains once this process is over, assuming the relationship is still a healthy one, is a steady emotional baseline where the person we fell in love with is now a kin who evokes an emotional response similar to that evoked by other kin—our parents, our siblings, or our children (assuming that nature got its way and reproduction happened). If things were rocky, separation may ensue. Rejection in the second phase has two consequences in the normal, nonpsychopathological case, where the lover opts for resignation: stage three occurs sooner, cutting short the duration of the second (psychotic) stage, and stage four still takes place without any modification to its duration of four to five years. What remains in this case is a fond memory of the person we fell in love with.

I am now going to introduce three imaginary protagonists named Jane, Ken, and Linda to illustrate the above four steps. Jane is a rookie real estate agent at a prestigious real estate firm in New York City. She graduated college with a degree in finance and worked as a trader at a major bank's bond trading desk for several years following graduation. Technology had been gradually reducing both the compensation and the number of traders. In addition, she had been feeling uninspired by the soulless and repetitive generation of profits that her job required. Spending her days facing a computer screen had also been wearing her down mentally. So when she was finally laid off along with several other traders, she almost felt relief. After five months of unemployment, and all the uncertainty and the financial strain with which it comes, she was finally hired by her current employer, one of the country's premier real estate agencies. She had always dreamt of being in a client-facing role and was, and still is, feeling infinitely grateful to her employer for giving her the opportunity. Starting on a new career path in the dog-eat-dog world of selling prime real estate property has not been easy. Her employer has steep sales goals and keeps close track of the performance of its recruits in their first three years at the firm. The performance goals are set to increase gradually over the course of this evaluative period. Continued employment is contingent upon meeting these preset goals month in and month out. She learned soon after starting on the job that the firm hires many rookie agents every year. Statistically, less than 10% of them make it past three years. Over the past year that she spent at the firm, she saw nearly half the people who had been hired around the same time as her being fired for failure to meet sales quotas. She has so far managed to stay on goal, but by a slim margin. The level of stress is already tremendous, and only getting worse as the performance goals are being ramped up. This is when she first meets Ken, a prospect who has expressed interest in one of the properties she is trying to sell. It is a starter home, not likely to be a big generator of commission.

Ken is a young cardiologist. He went through the grueling education and training requirements of medical school, which were followed by years of specialization as a fellow and further training as a resident. He has been practicing at a small hospital for several years, and his skills have earned him increasing levels peer recognition. His earnings have been steadily climbing after residency. They are about to jump by a significant amount as he has been engaged in negotiating a new position at a much bigger hospital in the city for several months. Both Jane and Ken are in their mid-thirties. Both are good-looking people. Neither has been particularly tempted by the prospect of parenthood so far. Jane has actually been averse to the idea and has been quite sure for a very long time that she wants nothing to do with motherhood. She has had an active and adventurous sex life since age of nineteen, and has had several boyfriends since then. Around the time that she started working for her real-estate firm, she moved in with her adored boyfriend of four years. As for Ken, he has always felt ambiguous towards the idea of fatherhood. Neither convinced that he wants it in his life nor certain that he does not, he has felt under increasing social pressure from both his parents and friends to find a marriage partner. He is also feeling some pressure centered on the idea of getting older. If he ends up getting married, he would prefer his wife to be a highly educated woman and one who would also be a successful professional. About six months ago, he broke up with his last girlfriend, another medical doctor, with whom he had enjoyed two great years. A week before he and Jane met for the first time, a friend introduced him to Linda, a twenty-seven-year-old graduate student working towards a PhD in statistics. She will most likely end up graduating, but realized many years ago that while she is good in statistics, she is far from being the best. Completing the requirements of her thesis research has been a struggle, so much so that she is no longer sure she wants to remain in the field. She has been feeling somewhat depressed lately and worried about other ways she might support herself. She is judged by most people to be very attractive. She has dreamt of being a mother since she was a little girl. She had a boyfriend throughout college, but their relationship was more emotional than sexual. In fact, she has always been more emotionally than sexually motived in intimate relationships.

If we think back to what we learned in chapter 3, the love circuitry of all three of our protagonists should be ready to fire at any time now. They have all been in an aroused emotional state for an extended time period. Their worn-out prefrontal cortex is most likely to loosen its inhibitory grip on the limbic system, their brain's emotional release center and home to its love circuitry, when stretched to its limits. Notice that their emotional arousal is gender-specific along the lines of the damsel-in-distress archetype. Both Jane and Linda have been feeling intense negative emotions due to existential anxiety related to an uncertain professional future along with high levels of work-related stress for Jane and some level of depression for Linda. Ken, on the other hand, has been riding an emotional high. His years of hard work are finally paying off. He is being recognized by peers professionally and is on the verge of getting a big promotion. He is reaching a point where he can comfortably play the ancestral role of the provider for a family. He is feeling exhilaration, pride, and hope. The future is looking bright. Notice also that he is feeling pressured by both his social entourage and his age. He would prefer not to have underage dependents by the time he retires. He is really starting to push against his age limit for remaining single. All three are ripe for an emotional surge, the most intense highs the human brain's reward system is capable of—the highs of falling in love.

Remembering the discussion of chapter 2, all three of our protagonists seem to have in-between brains with elements of both the female-typical and the male-typical brain. Judging from their level of education and subject matter focus, they all seem to have well-developed language and visuospatial processing centers. Both Jane's and Ken's brains have a slightly more pronounced tilt towards the systemizing brain, while Linda's seems to be somewhat more tilted towards the emphasizing brain. All three look their traditional gender and are perceived to be physically attractive. Both Ken and Linda are single. Jane has been in a loving relationship for more than four years. She is just as likely to fall in love—with her live-in boyfriend or someone else entirely—as Ken and Linda. Despite popular belief, the falling-in-love process has no consideration for exclusivity. Exclusivity only matters so far as it furthers our goals, the ones we talked about in the last chapter, namely ensuring paternity certainty, that the children get the food, shelter, and care they need until they can support themselves, and ensuring food, shelter, and care for ourselves in the process.

Jane and her boyfriend met online. She chose him carefully among many other men on the site. They are both lovers and best friends. They have felt attracted to each other sexually, physically, and emotionally since they met, but they never fell in love with each other. Falling in love is not sine qua non for a great intimate relationship. In fact, falling in love is highly likely to result in mismatching. Online dating offers us a far better process for starting an intimate relationship, allowing us to gather information, do some trial and error, and make a more rational decision about whom to go on dating for a longer time period. Notice also that our three protagonists differ in their feelings toward parenthood: Jane wants nothing to do with it, Ken feels ambiguous, and Linda has always dreamt of it. Despite these very diverging feelings, all three are capable of falling in love; our love circuitry has no regard for such subtleties. Ensuring reproduction—hence, the perpetuation of the species—will always have the highest priority. Of our three protagonists, Linda is the one for whom having a child in the near future would be the most unadvisable. She would be much better off giving herself time to graduate, figure out what career track she wants to take, and put in the four to five years of hard work required to jump-start any demanding career before embarking on parenthood. Since our love circuitry probably incorporates some of the most ancient elements of our brain, it is not designed for the demands of our modern lifestyles. It has no regard for any of our professional aspirations. One would also hope that all three realize they may be about to sign up for a legal contract lasting a minimum of eighteen years and a moral contract of a good twenty-five years or so (if not for life). These agreements carry a cost of half a million dollars per child, not to mention the staggering biological costs that women alone still have to endure. Indeed, today's estimates peg the financial cost of getting a child through high school at a quarter of a million dollars for a middle-class family. It can take another quarter of a million dollars to get them through the first year or two following college graduation. Our evolved love circuitry is not designed to uphold the best interest standard for those it ensnares. It could care less about our happiness. It is designed to accomplish one and only one thing: making sure that reproduction happens. It is up to our prefrontal cortex to bootstrap solutions to all the consequences of that event. Unfortunately for all those who were burnt by this process and are left wondering what led them to take the actions that they did, the enemy is within. We all have an innate tendency to seek agency in whatever befalls us and to seek fault outside. The culprit is not their culture, their parents, or their friends. It is not even the person they fell in love with. It is their brain's love circuitry.

Now, let's get back to our story. When Jane opened the door of the empty house she was attempting to sell and saw Ken's face smiling confidently for the first time, the thought that immediately flashed through her mind was *He's hot!* She smiled back and shook his hand with a professional, "Nice to meet you." As they toured the house, she learned a great deal about what he wanted and did not want in his starter house. He gave her an idea of the increased income he was in the process of negotiating. She could not help but notice that he was bragging a bit as he did so. He told her that he was thinking about buying some rental real estate in the future and, therefore, interested in finding an agent who he could trust for ongoing professional guidance. Since he expressed some reservations about this particular house, Jane offered to have him visit another property with her sometime the following week, to which he agreed. As often happened when driving back home at the end of a long workday, Jane's tired mind was lost in a state of daydreaming, hopping from topic to topic. She caught herself thinking about how handsome she found Ken. She immediately scoffed at herself for being silly. This was hardly the first time she had met an attractive prospect.

About a week before Ken and Jane met, a mutual friend had urged Ken and Linda to connect. They had briefly spoken over the phone and agreed to meet at an upscale bar downtown for happy hour. He approached her table and asked, "Linda?" She turned her head around and responded with a smile.

"Yes. Ken, I assume?" His first thought was *Wow, she looks amazing!* He caught himself reflexively turning his gaze away in embarrassment. He managed to control himself, looked back at her, smiled.

"Yes. It's great to meet you," he said. In ruminating about the encounter a few years later, he felt sure that it was at this precise moment that a switch was seemingly turned on in his head. He sat down beside her. They shared a pleasant conversation over drinks and appetizers. Eventually they found themselves agreeing to a second date over the weekend. In the days leading up to their second encounter, Ken caught himself several times during breaks at work or in the evenings at home mentally replaying in great detail parts of their interaction, especially those where they had both felt and expressed a pleasant emotion. These vivid flashbacks would cause him to mentally relive those moments, feeling the very same intense emotions, and to unconsciously smile as if she was there facing him again. These flashbacks had become both more frequent and more intense by the time they had sex for the first time several weeks later. The flashbacks were often complemented by fantasies of a sexual nature, to which he often masturbated. As he did so, he sometimes imagined that she was thinking in the very same way about him at that very instant, that she masturbated to thoughts about him in tandem, and that there was a telepathic sort of connection between them. The fantasies included a glamorized version of Linda: she was a top-notch statistician on her way to becoming a professional sensation, she had the kind heart of an angel, she was a classical beauty, she had strength, grit, wit, integrity, elegance, and

64

warmth, along with many other traits he found to be most charming in a woman. Scrutinizing the memory of their earliest dates several years later, Linda was certain that a switch had seemingly been turned on in her head when, during their fifth or sixth date together, he had told her how smitten he had become with her and how obsessively he kept thinking about her. They had had sex for the first time that night. Until then, she had felt attracted to him, but had never spent too much time thinking about him. This particular date and the ones that followed etched memories so deeply into her brain that she could recall the smallest details of some of their interactions along with the finest components of the emotions they had triggered many years down the road.

Meanwhile, Jane had shown several houses to Ken without enticing him to buy any of them. By then, his new job had become official. Since he was in no particular hurry to buy a house and since he expected to be quite busy as he sank into his new position and the responsibilities it entailed, he told her to contact him six months later to show him more properties. In that month and the months that followed, Jane fell behind on her performance goals at work. She had a very unsettling meeting with her supervisor, who made it quite clear to her that she would be given three more months to get herself back on track or the firm would have to terminate her employment. The couple of months that followed became a living hell for Jane. The stress was unbearable. She could hardly sleep. But her boyfriend's unrelenting support and encouragement throughout the ordeal kept her persevering. She felt more grateful to be with him than ever. By fighting tooth and nail, she managed to meet her goals before her deadline. She was elated, but also left quite shaken by the experience. It was then that she received the reminder from her customer relationship management system to call Ken again. The timing could not have been better, as she had a property in mind that she was sure he would love. Its price was somewhat north of the price range he had indicated, but she had developed quite a bit of skill in upselling by now and needed the commission to meet her near-term goals. As she saw him again for the first time in many months, she recalled how handsome she had found him when she first met him. Following the tour of the house, he agreed to meet again to execute documents towards the purchase. As they were walking in the parking lot, Jane mentioned that she was from Cleveland.

"No kidding—me too!" said Ken. "I spent the first ten years of my life in Cleveland, then my family moved to Virginia Beach." Jane had spent the first eight years of her life there before moving to Wisconsin. Both their parents were native Clevelanders. In fact, her mother and his father shared a common Polish ancestry. Ken wondered how they never came to meet each other in childhood. Thinking back about their earlier meetings several years later, Jane was indubitably persuaded that it was at this precise moment that a switch was seemingly turned on in her head. In the days that followed, she caught herself several times, during her drive from one property to another or during her commute from work to home, mentally replaying in great detail parts of their interaction, especially those where they had both felt and expressed a pleasant emotion. The flashbacks were so vivid that she felt the associated emotions as intensely as if he were right there with her. She felt nervous when he met her at the agency to review the documents. Once during the interaction, she felt her cheeks flush when he smiled at her. She felt mortified at the idea that she may be giving herself away. She started daydreaming about him more frequently afterwards. As time went on, her fantasies became both more intense and more sexually tinged. She felt conflicted about the fact that she felt that way toward a client, which was unprofessional, but even more so about the fact that she had a wonderful ongoing relationship with her boyfriend. She managed to remain professional with Ken throughout the closing process. They agreed to be in touch in about a year to have a discussion about the possibility of purchasing his first rental property. With this temporary separation, her daydreaming about him did not stop. Instead, it intensified over time. As she commuted back home on many Friday evenings following an exhausting workweek, she would repeatedly have a particular fantasy where she invited him for dinner after a professional meeting. During the dinner, they would each confess the passion they felt for each other. She had this image of Ken in which he was a male version of herself: he had a wonderful live-in girlfriend of many years, a medical doctor, whom he adored, but his feelings had grown so much for Jane that he could no longer bear to keep them to himself. They both remained cold-blooded throughout their confessions and agreed that they could enjoy each other's bodies without hurting their existing relationships. They proceeded to flirt and kiss each other. No longer able to contain their arousal, they decided to check into a hotel room, where they had passionate sex. She would return to reality at the end of her commute. She would get home in this aroused state, start to seduce her boyfriend, and have voluptuous sex with him instead. It seemed to her that, rather than hurting her relationship, her feelings for Ken had reinforced it by reigniting the passion within it. In the middle of sex with her boyfriend, she sometimes imagined that Ken was the one on top of her. This only made her wetter and the sex more enjoyable. Sometimes she came home so aroused by her daydreaming that she would go into the bedroom, tell her boyfriend that she needed a moment alone, shut the door closed, and masturbate while thinking about Ken. She imagined as she did so that he was thinking in the very same way about her at that very instant, that he masturbated to thoughts about her in tandem, and that there was a telepathic sort of connection between them. She pictured him in the most hyperbolic ways, attributing him all of the traits she found to be the most charming in a man. It seemed to her that her feelings for Ken would never go away, that they would last for a lifetime. Meanwhile, she had managed to remain on goal throughout that entire year. In fact, once a few of the really big sales she was in the process of executing would be completed, she would reach her goal for the entire evaluative period and would become a senior agent within the firm. Then the moment finally came when her customer relationship management software would give her the reminder that it was time to get back in touch with Ken. They agreed to meet two weeks later. By that time, she would go to bed thinking about him and would wake up to thoughts about him virtually every single day. The thoughts were compulsive, entirely spontaneous, and seemingly uncontrollable. She was obsessed with him. She had an intense longing to be near him and craved sex with him. They were meant for each other. She had to act. There was no other way. She became persuaded that she had to seduce him during that meeting into joining her for dinner and, later, for sex—just like in her recurring fantasy. She even booked a room in an upscale hotel for that night and told her boyfriend that she was showing a property out of town and would be staying the night. She bought a sexy, yet professional, new dress for the occasion and a new pair of high-heeled shoes to go with it. Two days before the meeting, Ken texted her to tell her that he had an important commitment following the meeting and that he could only spare half an hour for it. Her hopes of seducing him were dashed. She still put on her sexy dress and shoes and went in for a short professional session with Ken. In thinking about it later, she felt infinitely grateful to have been prevented from making a fool of herself. Indeed, as they both sat down around a table, she felt a sharp, almost visceral pain when she noticed the wedding ring on his finger. She let him speak as she recomposed herself. He briefly told her about his recent marriage to Linda, and she congratulated him on the occasion. She went to her hotel room alone and heartbroken. She could not sleep that night. She spent her commute the next day wondering about what had happened to her in the past year.

In the months that followed, she realized that she only rarely fantasized about Ken. Instead, she spent her time analyzing her past feelings. Occasionally, when she felt exhausted at the end of a workweek, she might still have a flashback of a pleasant interaction they had had—only, most of time, it was devoid of the sexual component. With the sight of that wedding ring, the switch in her head had suddenly been turned off. It was as if a bursting stream of water had poured over the fire that had consumed her for an entire year and almost entirely extinguished it in a split second. Her intense feelings for Ken were not meant to last a lifetime after all.

The mental switch that seems to go on for all three of our protagonists represents the moment that their brain's love circuitry (buried within the limbic or emotional-release center) decides to trigger a hormonal chain reaction to release a cocktail of neurochemicals into their brain and bloodstream. These neurochemicals loosen the inhibitory action of our prefrontal cortex on our brain's limbic system, which is now in charge. Our brain's inhibitory circuitry is a critical component of our ability to make sound judgements. This circuitry usually gets impaired over time in dementia

68

patients—this impairment is the reason for their socially inappropriate and impulsive behaviors. People who are in love are often reported to engage in inappropriate and impulsive behaviors, of which Jane's later behavior is a prime example. This is stage one of falling in love (the split-second rise in emotional intensity). Our brain's love circuitry triggers the falling-in-love process once it has determined that the person in front of us has passed most elements of its ancestral checklist for a long-term mate and the moment is ripe for us to reproduce. Research shows that men are more prone to fall in love at first sight. This happens to Ken in our story. This is because their ancestral checklist is mainly made of visual checks. Going back to what we learned in the last chapter, men are primarily looking for signs of fertility (i.e., youth) and health (i.e., beauty). We saw that behavioral displays of chastity and fidelity have also been deemed critical by ancestral men. Those men in whom ensuring paternity certainty is more strongly wired may be less prone to fall in love at first sight. Their love circuitry may want to run checks on the potential mate's sexual reputation and assess their tendency for promiscuity before impairing their ability to make sound judgements.

Women are also capable of falling in love at first sight. More typically, though, they fall in love with a delay compared to men. This is because the risk of being seduced and then abandoned has ancestrally loomed large for them. Pregnancy has traditionally been a massive and life-threatening cost to them—with no way to avoid it short of abstinence, in times that preceded birth control, meaning almost the entirety of human existence. And the costs do not stop at pregnancy. We discussed how earlier delivery has made the human baby more and more vulnerable, and delayed maturation has made the burden of raising a child heavier and heavier as our brains grew bigger. Choosing the right partner has been nothing short of a matter of life and death for ancestral women and their offspring. Ascertaining a potential mate's status among other men has topped women's ancestral checklist. It is a sign of his ability to acquire resources. The best metric for it today is socioeconomic status. The fact that Ken was a rising cardiologist probably influenced the love circuitry of both Linda and Jane. Had he been an equally good-looking janitor, neither may have fallen in love with him. Women's love circuitry also gets a high at the sight of expensive items, which is why young men dream of driving Maseratis. Outright display of resources is not quite as efficient as socioeconomic status, because it is very liable to manipulation: the guy could be a moron whose rich uncle gifted him the Maserati, or who borrowed it from his best friend for a while, or who even stole it from someone else. Even more important on women's ancestral checklist is the potential mate's capacity for generosity, meaning his willingness to share his resources with her and her offspring. A wealthy but stingy and callous mate is of little use. In popular parlance, this is typically referred to as emotional commitment, implying that the purse will follow the heart. It is this last check which may take a while, hence the delay in falling in love for women.

In our story, both Linda and Jane fell in love with a delay. Linda's delay was the shortest. She had ascertained Ken's socioeconomic status at their first date, but her brain's love circuitry needed his confession about him being smitten with her a few weeks into their relationship before it blinded her judgement. For Jane, the delay was much longer. It happened many months after meeting Ken. She has a stronger libido than Linda, so the visual check topped her list. She, too, knew about his socioeconomic status at the first meeting. Ken even purposely bragged about it. She fell in love when the degree of stress in her life went up a few notches higher and at the precise moment that Ken made a reference to Cleveland, their common birthplace, and to having common Polish roots. It is then that her brain's love circuitry made the determination that he represented the ideal compromise between inbreeding and outbreeding and triggered the falling-in-love process. Another reason for the longer delay for Jane may be the fact that she and Ken had a professional relationship marked by prolonged periods of restricted contact. It may also be that Jane's prefrontal cortex controls her limbic system more rigidly, making her emotions less prone to going off-kilter. Otherwise stated, her brain may inherently have a higher threshold in order to trigger the falling-in-love process. This view of falling in love fits in well within a psychological theory of mental disorders called the "diathesis-stress model," which asserts that if the combination of the diathesis, meaning predisposition or vulnerability (stemming from the genetics of the individual) and the stress (stemming from the individual's environment) exceeds a threshold, the person will develop a disorder. To the extent that falling in love is a psychosis—albeit one not usually considered to be a disorder—it will affect each of us with varying likelihood depending on our genetics and the specific life circumstances.

Following the date during which they first had sex together, Ken and Linda would each be counting the hours separating their next date. Linda, who had never been a very sexually motivated person, became voracious in bed. They sometimes had sex multiple times a day while spending a weekend together. Four months into the relationship, Ken proposed, and Linda immediately accepted. Their wedding took place six months after that. The fact that Linda was still a graduate student did not seem to bother either in the least. She was initially set to defend her thesis by the end of the summer, or about three months into their marriage. Her advisor suggested that she needed another six months to be ready. Her actual graduation ended up happening more than a year after the wedding. By that time, Linda was wearer with statistics than ever, and more enamored than ever with Ken. They had married so that they could have children together. She spent her days dreaming about him and feeling infinitely grateful to have him in her life. She became persuaded that she needed to prove her love by bearing his child. So she stopped taking her birth control a couple of months before graduating. A few weeks after graduation, she missed her period and bought a pregnancy test kit. When the result came out positive, she could hardly control her ecstasy. She constantly checked the time as she impatiently waited for Ken to return from work. When he did, she immediately told him about it. She added that she intended to take a year off to get through the pregnancy and to nurse her baby for a few months. It is at this moment that Ken felt a mental switch turn off. Rather than feeling happy about the news, he felt slightly irritated. Linda was surprised by his reaction. She had expected him to take her in his arms, kiss her, make love to her, and thank her. Instead, he told her that he had had a stressful day and needed to clear his head. He simply headed out telling her not to wait for him. What she experienced once alone was akin to a panic attack. A fear so deep overtook her that she could hardly breathe. Her whole

body was shaking. It took her a few hours to calm herself down. She took a
sleeping pill and went to bed. It was the next day, after Ken headed out to
work, that she realized that the emotional high she had been riding for a while
now was over. She spent the entire day dispirited. She no longer daydreamed
of Ken. Rather, she was absorbed in thoughts about her condition. In the
months that followed, Ken showed support and did his part in going to
doctor's visits with her. But he never put too much enthusiasm into it. He
seemed to be going through the motions. Meanwhile, the frequency of sex
steadily dropped. By the fifth month of pregnancy, Ken did not even want
to touch her anymore, which did not seem to bother Linda too much. She
did not particularly wish to engage in sexual activity anymore. She had been
feeling herself slipping slowly back into a depression. She reached the lowest
point shortly after giving birth. She started seeing a therapist and taking
antidepressants. She wanted to breastfeed her baby and was told that it would
be fine. Five months postpartum she started feeling a bit better, but sex was
the last thing she even wanted to talk about. Meanwhile, Ken's frustration
had reached its apex. He felt deceived and stuck in his marriage. Before
marrying Linda, he had understood her to be professionally ambitious. In
fact, he had at some point thought of her as a prodigy in statistics. In reality,
she was closer to being mediocre and did not seem to display the least bit of
professional aspiration. She was not even looking for job. They had not had
sex in almost a year. He often found himself fantasizing about an affair. Linda
was not the sex-hungry woman he had married. She had metamorphosed
into a prude. In addition, he was starting to realize that fatherhood was
far from being an exhilarating experience. The baby's constant crying and
neediness unnerved him. He cringed changing diapers so much that he was
running out of excuses to stay away from it. He started spending more and
more time at work so that he could spend less time at home. He began
to contemplate divorce, but felt bad about abandoning Linda and the baby.
Every now and then, he still had flashbacks of some of their happier moments
together during the time that he was in love with her. He noticed, however,
that those moments became less and less frequent since the announcement
of Linda's pregnancy, while the feelings of frustration kept growing. Their

marriage became so strained over time that they decided to file for a divorce about four years into it. Their child was about two years old then.

Stage two of the falling-in-love process is the phase where emotional intensity gets amplified over time. It is the core of what we usually call "being in love." Elvin Semrad is quoted by two of the former resident psychiatrists he supervised to have used the expression "the only socially acceptable psychosis" to describe this particular mental state. Frank Tallis, a psychiatrist and writer, simply calls it "a state of mental illness." The cocktail of hormones released into the bloodstream once the process is triggered contains, among others, biochemicals which make men and women more similar to each other: women become more interested in sex, and men become more interested in emotional connection. Ken, who only got to know Linda sexually while she was in love, was surprised to discover that she had become prudish shortly after her pregnancy was confirmed. In fact, a lower libido had always been Linda's baseline. Her libido was temporarily lifted while in love to enable reproduction. It may, however, have been brought below its baseline due to pregnancy and lactation. Pregnant women start to produce the hormone prolactin in preparation for lactation and go on to produce copious amounts of it as they lactate. Prolactin lowers libido and suppresses the effect of estradiol which is needed to maintain blood flow to female genitalia. Many lactating women report lower sex drive, vaginal dryness, and some level of vaginal atrophy. Their libido typically slowly recovers over the course of the year that follows delivery. It seems that nature really wants their full attention to be on their baby in the first year of the baby's life. In a way, prolactin acts as a natural contraceptive. It would be interesting for scientists to conduct studies in order to investigate whether some of the changes it induces are permanent.

The love cocktail also contains hormones that lift our mood and make us feel euphoric, hopeful, confident, and energized. Falling in love lifted Linda from her baseline of always being in a slightly depressed mood. She fell right back to it as soon as the process ended. Of course, Ken's lack of enthusiasm with respect to her pregnancy only exacerbated it, not to mention the voluntary unemployment that followed her graduation. Furthermore, pregnancy is a major hormonal event, a biochemical roller coaster, in addition to being a very traumatic one both physically and emotionally. Someone with a fragile neurochemical balance, such as Linda, may easily be tipped off equilibrium as a direct result of pregnancy. The condition is found to trigger a severe relapse in mental illness for many women with preexisting psychopathology. Some get diagnosed with a mental condition for the first time postpartum. In addition, a really interesting recent study found that pregnancy alters a first-time mother's (but not her husband's) brain structure, possibly in a permanent way (Hoekzema et al., 2017). It induces reductions in gray matter volume in regions subserving social cognition, especially those regions responding to the women's babies postpartum. This is accompanied by the strengthening of the white matter connections between the remaining neurons. The authors uncovered a correlation between volume changes in pregnancy with levels of maternal attachment. They further found the reductions in brain volume to endure for at least two years. The changes enhance a new mother's sensory abilities and theory of mind, presumably so that she may better deal with a physically impaired little being who is also devoid of speech. The structural changes were so dramatic that a computer program could tell between a woman who had been pregnant and one who had never been about 80% of the time. More research needs to be conducted to ascertain whether these brain changes are permanent and to investigate the effect of multiple pregnancies on women's brains. Many women have reported losing interest in sex altogether following the birth of a child or two—not just temporarily, but permanently. They essentially became asexual after having a child or two. Some found themselves losing all professional interest or losing all interest in past passions, such as a love for music or painting, after pregnancy. Could this be viewed a bit like the case of a biochemist experimenting in her lab to temporarily impair women's interest in sex and in outside endeavors so that the baby can get the full scale of her attention, but who ends up accidentally overdosing in certain cases so that some of them end up losing those interests permanently? Could Linda's new "mommy brain" be related to her lack of interest in finding a job despite going through the pain of earning a PhD degree? It remains an important task for social scientists to conduct epidemiological studies to quantitatively assess some of the life changes women experience as a direct result of pregnancy-induced brain alterations.

The psychotic element of falling in love consists of the detachment from the reality it causes. Ken's earlier belief that Linda was a prodigy in statistics was simply delusional. He never did any fact finding before marrying her. And had he done so while in love, he would most likely have dismissed it. It is the very nature of delusions to be immune to facts that directly disprove them. The idealization process by which we project the image of our dream mate onto the object of our love and delude ourselves into believing that the two are the same was identified two centuries ago by the French writer Stendhal in a psychological process he named "crystallization." Stendhal gives us a beautiful way to conceptualize the metaphor visually in the following manner: "In the salt mines, nearing the end of winter season, the miners will throw a leafless wintry bough into the abandoned workings. Two or three months later, through the effects of the waters saturated with salt which soak the bough and then let it dry as they recede, the miners find it covered with a shining deposit of crystals. The tiniest twigs no bigger than a tom-tit's claw are encrusted with an infinity of little crystals scintillating and dazzling. The original little bough is no longer recognizable; it has become a child's plaything very pretty to see. When the sun is shining and the air is perfectly dry the miners of Hallein seize the opportunity of offering these diamond-studded boughs to travelers preparing to go down the mine." While in love, Ken had a crystallized representation of Linda in his mind, which he had convinced himself to be her true self. Once the effect of the love cocktail ended and he slowly started to see her the way she had always been, he rationalized this discovery by thinking that she had changed. She was no longer the person he had met. Factually, the only thing that truly had changed was his own perception of her. While they were in love, both Jane and Ken had a delusion which is common in schizophrenics: they both believed that they could telepathically communicate with the object of their desire. This sort of magical thinking is oftentimes observed in psychosis, especially when it is associated with schizophrenia, and sometimes with bipolar disorder. All three protagonists experienced a detachment of their mood from reality. Our moods are our internal feedback system that tell us whether the actions we are taking in the world that surrounds us are effective or not in helping us to reach the goal or goals we are trying to achieve. Thus, we feel happy if are on track, sad if we are not, and anxious if we may be getting off track. Jane, for instance, was seriously off track when she slipped off her program goals at work and was given the warning that she may be let go soon unless she got herself back on track. Falling in love may be the evolved strategy which her worn-out brain utilized in its desperation for a breath of air, for a little joy in a reality that became too devoid of happiness, and too full of anxiety and stress. By detaching itself a little bit from its rough reality every now and then and immersing itself in fantasies, her mind found temporary refuge and the joy it craved in psychosis. Shakespeare may have been on the money when he wrote: "Love is merely a madness; and, I tell you, deserves as well a dark house and a whip as madmen do; and the reason why they are not so punish'd and cured is that the lunacy is so ordinary that the whippers are in love too." The euphoria that being in love produces along with the hypersexuality, grandiose delusions, obsessional thinking, and the heightened levels of confidence and energy are all quite characteristic of hypomania. The best model for the second stage of falling in love might be one in which the lover first enters a prodromal stage of very mild hypomania, goes on to develop more severe hypomania, and progresses on to a state of psychotic hypomania.

Both Jane and Ken found themselves compulsively absorbed in mentally reliving joyful past moments with the object of their passion through involuntary mental flashbacks (or precise replays of those moments retrieved from memory). Thus they felt the powerful emotions associated with those visual recollections. These flashbacks are quite similar to those experienced by people affected with post-traumatic stress disorder (PTSD). Those moments which are the most emotionally charged (whether pleasant or unpleasant) etch particularly deep memories into our brains. This may serve the evolutionary function of learning, or making sure that we avoid dangerous situations in the future and seek more of the pleasurable ones.

In the case of falling in love, they may serve the function of permanently associating the mental representation of our loved one with pleasure in our mind. Those flashbacks endure throughout the falling-in-love process into and beyond the fourth stage of exponential decay in emotional intensity. They are there to brighten our moods and make us persevere when the going gets rough, and it will inevitably get rough throughout pregnancy, delivery, and in the aftermath.

Both Jane and Ken obsessively masturbated to intrusive sexual thoughts about their loved one. As they did so, the social attachment system we have evolved kicked into high gear, producing copious amounts of oxytocin and vasopressin, the bonding hormones. Even though she knew very little of Ken and had little actual contact with him, Jane thus became attached to him, the same way that Ken and Linda became attached by repeatedly having sex together. The obsessive fantasizing and masturbating resulted in an amplification or snowballing of emotions, creating a sense of urgency within them and a compulsion to act—something within them was dictating that they must seduce their loved one into having sex with them now. And this is exactly what Jane did when she prepared to go into a professional meeting with her client with the full intension of luring him into the hotel room she had booked in advance. As Ken and Linda repeatedly had sex in the earlier part of their relationship, they not only became more strongly bonded to each other, but they also increased their chances of conceiving (if modern contraceptives were not there to interfere with nature's course). It is easy to conceive that the duration of this phase would vary from person to person.

Those whose neurochemical makeup is easily thrust off-kilter might reach the psychotic stage sooner than those endowed with more rigidly controlled limbic system. In our story, Ken started to have psychotic delusions within a few weeks of meeting Linda while Jane reached that stage about a year after first meeting Ken. Linda and Ken stayed in love longer (both for about two years) while Jane was only in love for about a year before her hopes were dashed. If Linda had become pregnant within a couple of months of meeting Ken, they may have stayed in love for a much shorter period of time. Had Jane's efforts been cut short within a few weeks of falling in love, her mental state might have been even shorter-lived. Bombarding us with this rich cocktail of hormones and neurotransmitters has to be a metabolically expensive process. Furthermore, the break from reality induced by psychosis is inherently dangerous, as it may impede our ability to fight for our own survival. Our brains cannot possibly maintain us in this state for very long. A compromise must be made between giving us an interval of time which is long enough to actively court our loved one, seduce them, and repeatedly have sex with them until conception occurs, while being short enough to ensure that we are not cut off from reality for too long. We have to be alive and connected to reality to care for the baby, after all. My guess is that our brains could maintain this state of being in love anywhere from a few months to a few years. John Money suggests that "a strong pair-bond maintains itself at its level of highest passion typically for a maximum of two to three years," which, he adds, "may be construed as nature's guarantee that a pregnancy will ensue." A scientific study to assess both the range and the average duration of this process would be quite interesting, albeit challenging and costly.

Stage three of falling in love is just as brief as stage one. It lasts for a fraction of a second and consists of the cessation of the biochemical chain reaction which was set into motion in the previous step. All it took in Jane's case was the sight of the wedding ring on Ken's finger. Her brain's love circuitry got the message that his resources were committed elsewhere; she should, therefore, not be wasting any of her time and effort on him. She felt a sharp pain when that happened, as if afflicted by a piercing knife, a pain actually caused by the sudden withdrawal of the powerful cocktail of feel-good biochemicals, similar to the sudden withdrawal symptoms an addict might experience. For Ken, being in love came to an abrupt end with his wife's announcement of her pregnancy. Nature had done its job and there was no need to continue the metabolically expensive process of powering on the biochemical chain reaction of being love. It is possible that Linda's brain stopped the process earlier than Ken's, at the time it got a convincing hormonal feedback from her reproductive system, when her uterus and ovaries started signaling successful fertilization. Ken's negative reaction to the announcement of her pregnancy may have merely brought it to light.

Stage three results in promptly and substantially reduced emotional intensity—we are talking about a drop of 90% or so. Stage four consists of a slow and exponential decay of the remaining emotional intensity to baseline over a period of four to five years at most. The function of this echo is to maintain pair-bonding strong enough to get a couple through the pregnancy, delivery, and the critical first three to four years of their baby's life. What this means in practice is that the mental flashbacks we experience while in love do not immediately stop once and for all at stage three. They just occur much less frequently from that point forward, and with ever-decreasing frequency over a period of a few years. For both ancestral and modern women, pregnancy was, and still is, a major state of physical disability. In all likelihood, our ancestral foremothers also breastfed their babies for longer time periods, presumably three to four years. This meant that both they and their baby critically depended on male provisioning for that time period. Abandonment by the male past this stage may have had less lethal consequences for the child. The abandoned woman may have been able to leave the child in the care of grandma or other female relatives at this stage to go about gathering food. The evidence for the longer lactation period can still be seen in modern India, where mothers commonly breastfeed their children until two to three years of age. In many preliterate communities extant today, babies tend to be spaced by about four years. This is because prolactin (the lactating hormone) may be acting as a natural contraceptive. Levels of prolactin steadily rise during pregnancy and continue to be high in lactating mothers, leading to significantly lowered libido and, sometimes, painful intercourse due to some level of urogenital atrophy. If our *Homo erectus* foremother had not been abandoned by her male partner by then, she and her mate may have settled into the companionate phase of their relationship, sustained not by passion, but by feelings of security and by pragmatic considerations associated with the pooling of resources. The emotional baseline they would have reached then is one that applied to any other kin relationships they might have had. Their brains would have sheltered their mental representation from the overt criticism and judgement we typically reserve for strangers. The duration of the exponential decay that takes place in stage four of falling in love most likely varies from individual to individual. The ability to attach to others and the strength of the attachment is also variable from person to person. Women, on average, tend to have a stronger potential for attachment than men. This directly follows from the understanding that attachment to her intimate partner was more directly tied to an ancestral woman's survival than to that of an ancestral man's. The strength of their attachment system might, in part, explain why many battered women today will keep going back to their abuser.

In our story, stage four lasted about three years for Ken and Linda, getting them through Linda's pregnancy and roughly the first two years of their baby's life, at which time they sought divorce. One can easily imagine that this process would be much harder on Linda than on Ken, since she was still unemployed. She can be grateful to our modern justice system, which will grant her alimony for a while, together with the relief of child support and shared custody until her child reaches the age of eighteen. Our foremothers could only rely on the much less efficient honor system enforced by their tribe. It is a shame that a large fraction of modern fathers will fail to pay child support and cut off all ties with their children. Their exes are left to rely on their own efforts and the help of the modern welfare system. The latter is far superior to begging within the family, an unfortunate circumstance that befell many of our foremothers. Ken discovered along the way that he did not really enjoy being a father. Linda wanted to be a mother and she, at the very least, got to realize that wish. And of course, despite everything, their baby got a life. Nature got to have its way. Jane may be the luckiest one in our story. If you recall, she was very averse to the idea of motherhood from an early age. Had she become pregnant while in the state of psychosis we call "being in love," the consequences may have been far more unpleasant. Instead, falling in love reignited the sexual passion within her existing relationship with her boyfriend, making it stronger. Thanks to this luck, she was prevented from the harm she might have inflicted upon herself. She got to keep fond memories of Ken. In addition, the two of them continued working together as Ken fulfilled his ambition of investing in rental real estate.

5

Biochemistry and Neuroanatomy of Love

In the mid-1980s, Michael Liebowitz noticed the similarities between the emotional highs we feel when we are in love, those induced by psychoactive drugs such as cocaine or amphetamines, those felt in psychiatric conditions such as mania, and those that result from peak experiences as diverse as climbing Mount Everest, getting a dream promotion, building a financial empire, creating a work of art, or the mystical experience of feeling one with the universe. The commonality between these drug- and nondrug-induced experiences is their operation on similar brain circuitry by way of similar biochemical changes within us. We are all born with cortico-limbic programs that are genetically preset to run certain biochemical chain reactions within given brain circuitry when triggered by certain environmental cues in certain environmental circumstances. For the sake of parsimony, nature seems to co-opt the same circuitry and programs to differing environmental stimuli.

In chapter 4, I proposed a model of stage two of falling in love—the period of highest passion—in which the lover first enters a prodromal state of mild hypomania, goes on to develop more severe hypomania, and progresses on to psychotic hypomania. In psychiatry, *prodrome* refers to the period characterized by changes to mental state compared to the person's premorbid baseline, including changes in perception, beliefs, cognition, mood, affect, and behavior before becoming psychotic. The symptoms are preclinical and mild enough that they would not warrant a psychiatric diagnosis. Biochemical snowballing over time leads to somewhat more noticeable changes in the individual as he or she slowly progresses into a state of more pronounced hypomania. A hypomanic person may exhibit changes such as decreased need for sleep, increased energy, talkativeness, increased self-confidence, elevated sexuality, exalted mood (or euphoria), disinhibition, pressured speech and activity, and flight of ideas. No functional impairment is present at this stage. In fact, many well-known people in the creative fields have credited hypomania with giving them their imaginative edge. Kay Redfield Jamison proposes a direct link between hypomania and artistic genius. One can easily understand how euphoria along with increased levels of energy, activity, self-confidence, sociability, and talkativeness could help in a variety of competitive and challenging professional situations including direct sales, entrepreneurship, and sports. Add in hypersexuality, and the same mental process could help in courtship. In fact, any situation that requires a tremendous amount of focus and effort for a brief time period and offers a big payoff for the winner would favor the hypomanic person over his or her competitors, assuming that they are roughly equal in every other way. That such a person may sink into clinical depression in the aftermath would not matter at all since they already secured their big win. This may be the fundamental reason why such a debilitating mental illness as bipolar disorder has not been entirely wiped out by evolution. When coupled with a high degree of a particular talent or ability and a good dose of luck, it has the potential to turn the afflicted individual into a millionaire movie star or rock star or Nobel Prize–winning scientist or world-famous painter or renowned composer. In the story in chapter 4, Jane's brain may have specifically called on the falling-in-love process in the middle of her three-year evaluative period at her firm to help her win the battle when her monthly sales goals and stress levels were significantly increased, and she came close to losing her job as a result. It may have resorted to this evolved mental strategy to take advantage of the increased focus, energy, and activity that comes with being hypomanic. It took an enormous risk by also making her vulnerable to getting herself impregnated despite her lifelong aversion for it. But the stakes became so high that it judged the risk worth taking. The process can work in reverse, too. In Ken's case, the euphoric and self-confident mental state induced by winning the battle of climbing the professional ladder may have triggered its mental kin of falling in love when he was exposed to the right environmental stimulus—meaning the sight of a good reproductive candidate. Otherwise stated, a process we may have originally evolved to win in courtship over rivals has been co-opted over time to help us win in other competitive or creative situations. Mental illness only happens when the process is dialed up beyond a certain threshold or in situations that really do not warrant it.

This is in line with new research findings in genetics showing that mental illness represents the extreme of traits that come on a continuum and are each coded by hundreds or even thousands of single nucleotide mutations, each contributing a small percentage effect to the overall trait. Whether we are neurotypical or mentally ill depends on how many of these point mutations our DNA contains and how many are being expressed in any given moment. By the same token, people who have too many of these point mutations and may have therefore been diagnosed with mental illness at some point in life, being endowed with a more mercurial biochemistry, may also be prone to falling in love both more frequently and more quickly than those people who have fewer of them. The progression from the prodromal stage of mild hypomania to gradually more severe hypomania only represents an escalation of the same process over time, the difference being in the degree of intensity of symptoms. The condition may still be subclinical at this stage. Mildly psychotic elements may have started to form. As more time passes, and the lover still has not reached a resolution, he or she may enter a mental state resembling psychotic hypomania. Being in love represents a *functional* state of psychotic hypomania, meaning the individual is not so severely cut off from reality that they cannot perform their job or engage in other routine activities. Psychosis is characterized as disruptions to a person's thoughts and perceptions that make it difficult for them to recognize what is real and what is not. These disruptions are often experienced as sensing and believing things that are not real or having strange, persistent thoughts, behaviors, and emotions. Sensing (i.e., seeing, hearing, smelling, tasting, or feeling) things that are not real is called *hallucinating*. Holding on to certain beliefs despite factual evidence disproving them is called *being delusional*. While in love, persistent thoughts, consisting of mental flashbacks of pleasurable moments with the loved one, are obsessively and compulsively retrieved from memory and relived as if real. The visual memories bring along the associated emotional memories, and the lover feels them as if the object of their passion was nearby in a process akin to hallucinating. For a comparison, although the initial hallucinogenic effects of the psychedelic drug LSD only last for approximately eight hours, users can experience LSD flashbacks for years after they have stopped using the drug. The intrusive—although pleasurable—flashbacks experienced during stage two, the stage of highest passion in the falling-in-love process, also continue on throughout the four-to-five-year duration of the last stage, the exponential decay in passion. The crystallization process in which the lover attributes grandiose traits to their loved one, which they do not truly possess in reality, works by way of delusions. Each of the scintillating pieces of crystal attached to the bough the miners left in the salt mine during wintry season in Stendhal's anecdote represents a particular delusion the lover came to attach to his or her mental representation of their loved one. This mental image, studded in diamonds, shines bright in the lover's daydreams. In addition, lovers may come to have strange thoughts, magical in nature, such as believing that they can telepathically communicate with their loved one or feeling that they are united to them mentally or emotionally. The feeling of being one with one's surroundings—with all the people in the world, with nature, and with the universe—is quite characteristic of the peak mystical experiences people report during deeply felt religious rituals or a good LSD trip. Excessive religiosity and magical thinking are hallmarks of schizophrenia. Hallucinations, delusions, magical thinking, obsessions, and compulsions are all symptoms of psychosis, whether it is associated with the process of falling in love or experienced during a religious event or manifested in aggravated phases of psychiatric conditions such as mania, schizophrenia, and clinical depression, or induced by way of recreational drugs such as LSD or cocaine.

Again, it follows from all this discussion that we may have originally evolved psychosis to aid in reproduction. No wonder then that falling in love could be called "the only socially acceptable psychosis." It helps to perpetuate the species. One may argue here that creative psychosis can also get a favorable social nod. Indeed, the wild antics and eccentric lifestyles of many rock stars or other celebrities are not just tolerated, sometimes within extremely conservative cultures, but even celebrated and rewarded—when they would be spurned or even punished in the average person. A deeper dive into the lives of such celebrities often reveals psychiatric conditions. Psychosis, then, is either applauded or despised depending on whether it benefits the species or harms it. When it enables reproduction or fuels creativity that results in a beneficial invention or art, it is celebrated. But, when an individual becomes such a burden on everyone else in their family that they start to weigh them all down with their wild antics in order to extort as much for themselves as they possibly can—because the ultimate goal in that case is to obtain more help than one really needs—it is reviled. In the most severe cases, the family gives up on the afflicted person, who, in many instances, ends up spending the rest of their life between mental institutions, the prison system, and homeless shelters. Psychosis is the common thread between the creative genius and the mental ward patient, between humongous success and extreme misery, between beneficial creation and the worst extortion scheme, the highs induced by cocaine use and the unbearable pain of withdrawal, and between the lover's bliss and plight of the heartbroken. It is a double-edged sword: the sharpest kind. And, so long as humanity gains more from it than it loses, evolution will keep it along, no matter how much harm it can cause in the lives of specific individuals.

Exogenous drugs work in our body by altering the rate at which naturally occurring biological reactions proceed, not by creating new reactions. Stimulant drugs such as amphetamine and cocaine act on the human brain's reward pathway by increasing the production of excitatory biochemicals, in the same way that interacting with the person we fell in love with does. Cocaine is mainly used recreationally as a euphoriant. Used in therapeutic doses, amphetamine not only induces elevated mood, but also mimics some of the other effects of falling in love, such as increases in sex drive, wakefulness, attention, and focus. If continuously administered over a period of days or repeatedly used in escalating doses, it can cause psychosis, the same way that falling in love does when allowed to snowball over time. Psychedelic drugs such as LSD, ecstasy, or mescaline are known to induce altered thoughts, feelings, and perceptions similar to those experienced in certain dreams, mystical states, psychosis, or during the more intense daydreams of the lover. LSD can cause hallucinations and illusions colloquially known as "trips." Good trips are described in ways that much resemble the experience of falling in love: they are stimulating and pleasurable, and typically involve feeling as if one is floating and disconnected from reality, joyfulness, decreased inhibitions, and the belief that one has extreme mental clarity or superpowers, all of which are also reminiscent of the symptoms of manic psychosis. Bad trips involve more negative feelings such as anxiety, irritation, and paranoia and are closer to schizophrenic psychosis. This last mental state may induce feelings that are reminiscent of what a person in love may feel while in the throes of jealousy. It has been suggested that women may have evolved schizophrenia in their quest to weed out incompetent or absent providers. Could it be that men may have coevolved it in their obsession to ensure paternity certainty? This may, in part, explain why the condition tends to occur at similar rates in both sexes. It follows that mania, schizophrenia, and psychosis may have originally been evolved to play helpful roles in reproduction, at least in their milder nonpsychopathologic forms. Their extreme manifestations, then, represent disorders of the attraction system. Opioids such as heroin, morphine, codeine, and thebaine, on the other hand, produce mild euphoria, reduce pain, and alleviate anxiety by suppressing activity in a brain area called the locus coeruleus which appears to be involved in producing panic attacks and separation anxiety. Their impact is similar to that of endorphins, the feel-good hormones that our brain's attachment system produces, usually in conjunction with two close biochemicals called *oxytocin* and *vasopressin*, which also modulate pair-bonding. In fact, the term *endorphin* consists of two parts: *endo-* and *-orphin*. These are short forms of the words *endogenous* and *morphine*, intended to mean "a morphine-like substance originating from within the body." Should we be concerned knowing that our brains can produce biochemicals that can affect us the same way, or quite possibly in a far more potent way, as heroine?

The main difference between endogenous and synthetic narcotics seems to be the rapidity with which the former are cleared from our systems, while the latter tend to linger around for a bit longer. Copious amounts of these hormones are produced when lovers cuddle, have sex with, or masturbate while fantasizing about the object of their passion. They develop attachment for—or dependency on—the object of their passion as they do so. The effect of opioids is a particularly good model for the way that companionate love affects us. The latter is the baseline that remains in long-term intimate relationships once all four stages of falling in love are through. It is dominated by feelings of security, calm, peacefulness, and a state of mild euphoria. Any threat to the relationship can induce anxiety, fear, and pain, especially in those individuals who are prone to separation anxiety and panic attacks. Their addiction to their long-term partner is quite similar to that of an addict to heroin. Indeed, heroin can induce both profound degrees of tolerance and physical dependence. Tolerance occurs when more and more of the drug is required to achieve the same effects. Within the context of an intimate relationship, this can be compared to the slight dullness that envelopes a couple's interactions in the companionate phase. With physical dependence, the body adapts to the presence of the drug, and withdrawal symptoms occur if use is reduced abruptly. Symptoms may include hot flashes or cold sweats, chills, malaise, nausea, vomiting, cramps, and severe muscle and bone aches. All of these symptoms can be present in panic attacks that some people may experience when suddenly abandoned by their intimate partner—in addition to anxiety and depression, also common following a breakup. It follows that separation anxiety, panic attacks, and depression along with their psychosomatic symptoms (meaning the feeling of physical pain that sometimes results from extreme psychological distress) may well have been evolved to aid in reproduction, at least in their milder forms. Their more extreme manifestations, then, represent disorders of the attachment system. Since women have traditionally carried almost all of the burden of bearing and rearing children, they are the ones who would have benefited most from attachment to a long-term male provider. It should, therefore, not be surprising that they would have ended up evolving stronger attachment systems than men, which would make them more prone to the aforementioned mental illnesses which represent the extreme manifestation of the traits associated with it. Reality does, indeed, reflect this, since two thirds of the people affected by panic attacks and other anxiety disorders (including separation anxiety), depression, and psychosomatic disorders happen to be women.

It is interesting to make a slight digression here and note how the notion of dependence is inextricably intertwined with that of long-term or companionate love that we typically feel for our reproductive partner, blood relatives, and, to a lesser degree, our extended family and friends. The very same attachment system and biochemistry is involved in forging and maintaining these bonds. The higher the level of dependence, the stronger the bond, which we like to simply call love. Thus, the bond between a mother and her baby is the strongest possible within the realm of human experience, since the human baby is so desperately dependent on his or her mother—especially earlier on. The bond between the baby and their father could be next, but it is highly variable in degree since many fathers have traditionally had little (or no) involvement with their children, especially in the early years, although this is changing towards more involved fathers today. Children have still relied on their father's provisioning and worldly skills, and, therefore, would still be expected to develop strong attachment to them. Women have traditionally almost entirely relied on men for their survival (something that has changed quite a bit already and is continuing to evolve). Therefore, their attachment to—and dependence on—their reproductive partner is probably very high (and variable within a wide range given the large biological differences among women). Men have depended on women to bear and rear their children and would, therefore, be expected to develop dependency on them. One can expect their dependency (or attachment or love) to be highest at the beginning and wain more quickly over time since it has been possible for them to spread their reproductive contribution (which has consisted primarily of economic resources) between several women to achieve genetic diversity and ensure that their DNA could survive despite changing environments. This may sound sexist, but it is really more a statement of fact: women would have benefited in the same way from having children with different men and probably did so away from the public eye (rather than openly like men in polygamous societies) because they would have risked losing the provisioning by a single male, especially if it was from a talented and reliable one. Our attachment to our siblings is typically less than the attachment we feel for our parents; siblings routinely help each other, but do not usually develop total reliance on one another. Then comes the strength of our bond to our extended family and to our friends with whom we tend to establish relationships based on reciprocal help. The help we extend to close blood relatives is, in most cases, not reciprocal. Some people in the family are irredeemably mostly takers, and some are mostly givers. Is the word "love," then, the scintillating lacquer we use to cover up our underlying dependency on another person?

In order to better explain the impact of falling in love, other peak experiences, mental illness, and psychoactive drugs on our brains, I need to briefly describe how our nervous systems are built and how they work. The nervous system is made of the central nervous system consisting of the brain and spinal cord and the peripheral nervous system which includes the rest of the nerves and sense organs. Its function is to regulate information—meaning to receive, transmit, and process it. The nerve cell, or neuron, is the building block of the nervous system. It consists of a cell body equipped with many short tentacles called *dendrites*, which bring information in, and a long, cord-like extension called an *axon*, which transmits information out to the next cell. The nerves that run throughout our bodies consist of groups of axons banded together. A synapse is the minute space separating the axon of a signal-passing neuron (the presynaptic neuron) and one of the dendrites of the target neuron (the postsynaptic neuron). An electrical signal travels through the length of one cell to stimulate the release of a packet of chemical messengers called *monoamine neurotransmitters* when it reaches the end of its axon. These neurotransmitters are transported out of the axon into the synapse and stimulate the receptors on the edge of the next cell. This, in turn, induces the receiving cell to send an electrical signal along its length. *Reuptake* is the reabsorption by a presynaptic neuron of monoamine neurotransmitters that it has just secreted and released. It happens when excessive concentrations of neurotransmitters are detected in the synaptic plasma. Monoamine transporters (MATs) are protein structures that reside just outside of the presynaptic cleft and serve to transport or mop up the neurotransmitters from the synapse back into the neuron that emitted them. Of the several dozens of neurotransmitters that are known to exist, three play a particularly important role for our topic: norepinephrine, dopamine, and serotonin. There is an MAT associated with the transport and reuptake of each of these monoamine neurotransmitters: the dopamine transporter DAT, the norepinephrine transporter NET, and the serotonin transporter SERT.

MATs are commonly associated with drugs used to treat mental disorders, as well as recreational drugs—a line that can become quite blurred at times. For instance, clinical depression involves dysfunction of serotonin and norepinephrine circuits. The most widely used antidepressants are called selective serotonin reuptake inhibitors (SSRIs) and include drugs such as Prozac. These drugs inhibit the reuptake of serotonin from the extracellular space into the synaptic terminal by selectively inhibiting SERT. The tricyclic antidepressants selectively inhibit NET, leading to increased synaptic concentrations of norepinephrine, and work well in some cases of depression. Serotonin-norepinephrine reuptake inhibitors (SNRIs) act by blocking both SERT and NET and are also used to treat both depression and anxiety disorders. Schizophrenia is associated with dysfunction of dopamine and norepinephrine circuits. The serendipitous discovery of chlorpromazine in the 1950s, the first antipsychotic drug, is considered to be one of the major breakthroughs in psychiatry. It is used to treat psychotic symptoms of schizophrenia, bipolar disorder, and stimulant-induced psychosis, and works by blocking dopamine receptors on the target cells. It follows that high levels of dopamine may play a role in psychosis. Furthermore, both stimulant drugs (such as amphetamine and cocaine) and psychedelic drugs (such as LSD

90

and ecstasy) exert their influence in part by their interaction with MATs, by blocking the transporters from mopping up norepinephrine, dopamine, and serotonin from the synapse. Lithium carbonate is an antimanic agent used to treat bipolar disorder. It acts by decreasing norepinephrine release and by increasing serotonin synthesis. It follows that mania is characterized by high levels of norepinephrine and low levels of serotonin. Lithium carbonate is also effective in countering the effect of psychostimulants. The level of norepinephrine is usually positively correlated with that of dopamine and negatively correlated with levels of serotonin.

I proposed a model for stage two of falling in love in which the lover's mental state progresses from mild hypomania to psychotic hypomania over time. A good recreational drug cocktail to mimic the effect of being in love might be one with a dose of the stimulant amphetamine mixed with a dose of the psychedelic LSD and the opioid morphine. Give yourself repeated shots of this powerful drug cocktail and, over time, you may get close to feeling what people in love can feel at the acme of their passion. The synthetic drug model of falling in love would thus be one in which a person takes increasingly higher doses of this cocktail over a period of time. The cocktail's makeup would vary along with dosage as a function of time from mostly having amphetamine at the beginning, then incorporating some LSD and a minute dose of morphine, progressing to equally high doses of amphetamine, LSD, and morphine, and consisting mostly of morphine towards the end. Did I scare you enough? Based on the discussion of the preceding paragraph, the endogenous (or biological) drug cocktail our bodies produce when we are in love would include a high dose of norepinephrine and a low dose of serotonin (to mimic hypomania), a high dose of dopamine (to reinforce hypomania and add in the psychotic element), a high dose of testosterone (for increased libido), high doses of oxytocin, vasopressin, and endorphins (to ensure bonding or, in other words, induce dependence), and high levels of the nerve growth factor (NGF) to grow nerve cells in the hippocampus for memories involved in the compulsive flashbacks. Other hormones may be involved such as cortisol and epinephrine. Next, I want to go into the detail of the role that each of these biochemicals plays when we fall in love.

Norepinephrine

Norepinephrine (also called noradrenaline) is made in both the brain—where it functions as a neurotransmitter, passing information between neurons of the central nervous system—and in the adrenal glands to send hormonal signals to target tissue in the rest of the body. Its release is lowest during sleep, rises during wakefulness, and reaches much higher levels during situations of stress or danger—where our brains determine that a fight-or-flight response is necessary on our part. In the brain, norepinephrine increases arousal and alertness, promotes vigilance, enhances formation and retrieval of memory, and focuses attention; it also increases restlessness and anxiety. In the rest of the body, norepinephrine dilates pupils, increases heart rate and blood pressure, triggers the release of glucose from energy stores, increases blood flow to skeletal muscle, reduces blood flow to the gastrointestinal system, and inhibits voiding of the bladder and gastrointestinal motility. The critical role that norepinephrine plays in wakefulness and action is illustrated by the consequences of loss of norepinephrine-secreting neurons in the sympathetic nervous system in diseases such as Parkinson's and diabetes. Severely affected people cannot stand for even a few seconds without fainting due to a strong reduction in heart rate and an extreme drop in resting blood pressure.

The main point of production of norepinephrine in the brain is the locus coeruleus, which I mentioned earlier in relation to its role in panic attacks. This brain structure is quite small in absolute terms—in primates, it is estimated to contain around 15,000 neurons, representing less than one millionth of the neurons in the brain—but it sends projections to every major part of the brain and to the spinal cord. Its activity levels shoot up in response to unpleasant stimuli such as pain, difficulty breathing, and uncomfortable heat or cold. Extremely unpleasant states such as intense fear or intense pain are associated with very high levels of locus coeruleus activity. Locus coeruleus is also the part of the brain thought to be involved in panic attacks and separation anxiety. It is also quite active in the brains of people in love. There is a positive correlation between levels of norepinephrine and levels of dopamine. Stimulant drugs, both recreational and medicinal, whose primary effect is to increase dopamine levels in the brain, also increase brain levels of norepinephrine, making their relative contribution difficult to disentangle.

Levels of norepinephrine are elevated when we are in love just as they are when we are under extreme duress. One reason that our brains cannot possibly maintain us in the state of being in love forever is the physiologic cost we incur as a result over time. In physiology, stress is defined as any situation that threatens the continued stability of the body and its functions. Chronic stress, if continued for a long time, can damage many parts of the body. A significant part of the damage is due to the effects of sustained norepinephrine release—damage caused by norepinephrine's general function of directing resources away from maintenance, regeneration, and reproduction, and toward systems that are required for active movement. The whole idea is to focus all of our attention and allocate most bodily resources to action in the immediate present when we are in a situation requiring the fight-or-flight response; if we do not survive this very moment, we will not be here to see the future. Therefore, critical parts of the body are deprived of needed resources for optimal function and our future needs are entirely neglected. Physiologically, the consequences can include slowing of growth (in children), sleeplessness, loss of libido, gastrointestinal problems, impaired disease resistance, slower rates of injury healing, depression, and increased vulnerability to addiction (Chrousos, 2009). In addition, future needs are sacrificed for the sake of surviving the present. This is the reason why panicked people often make rash decisions—including poor financial decisions—that hurt them in the long run. Higher norepinephrine levels may be one major reason why being in love makes us blind, inspires us with a sense of urgency to act (meaning, to get ourselves pregnant or to impregnate someone else), and, in a way, dumbs us down. Those rash decisions, which appear to be absolute musts in the immediate present, are highly likely to lead to regret in the long run.

Dopamine

Dopamine, too, functions as both a neurotransmitter in the brain and a hormone in the body. It is also synthesized in plants and most animals. Dopamine cannot cross the blood-brain barrier, so its synthesis and functions in peripheral tissue are to a large degree independent of its synthesis and functions in the brain. In the body's peripheral systems, it is synthesized locally and exerts its effects near the cells that release it. In blood vessels, it inhibits norepinephrine release and acts as a vasodilator in direct response to increased heart rate and blood pressure during the fight-or-flight response; in the kidneys, it increases sodium excretion and urine output; in the pancreas, it reduces insulin production; in the digestive system, it reduces gastrointestinal motility and protects intestinal mucosa; and in the immune system, it reduces the activity of lymphocytes. The brain includes several distinct dopamine pathways, one of which plays a major role in the motivational component of reward-motivated behavior. Other brain-dopamine pathways are involved in motor control and in controlling the release of various hormones.

In the brain, dopamine-producing neurons are comparatively few in number (a total of around 400,000) and their cell bodies are confined in groups to a few relatively small brain areas. Their axons, however, project to many other brain areas and exert powerful effects on their targets. The basal ganglia consist of a group of subcortical nuclei which are situated at the base of the forebrain and top of the midbrain. Two areas of the basal ganglia—the substantia nigra and the ventral tegmental area (VTA)—are responsible for most dopamine production in the brain (about 85% of the total). The basal ganglia are thought to play a central role in action selection, meaning the selection of a particular action to engage in when faced with a situation calling for several possible behaviors. The level of dopamine sets the threshold for initiating actions. High levels of dopamine lead to high levels of motor activity and impulsive behavior—similar to those observed in hypomania, drug-induced highs, or during stage two of falling in love. Low levels of dopamine lead to torpor and slowed reactions—similar, in extreme cases, to those observed in people who take antipsychotics such as chlorpromazine, as well as those afflicted with Parkinson's disease. Dopamine's other important role is in helping us to learn. When an action is followed by an increase in dopamine activity, the basal ganglia circuitry is altered in a way that makes the same response easier to evoke when similar situations arise in the future.

While the VTA guides action selection by inducing varying levels of pleasure, the substantia nigra plays the next executive function of choosing particular sets of muscles to activate in order to carry out the action. The role of determining the remaining details of the chosen action is relegated to other brain systems. The most prominent group of VTA dopaminergic neurons projects to the prefrontal cortex via the mesocortical pathway and another smaller group projects to the nucleus accumbens via the mesolimbic pathway. These two pathways play a central role in reward and other aspects of motivation. For motor aspects of action selection, both the VTA and the substantia nigra send dopamine signals to an area of the basal ganglia called the corpus striatum.

Pleasure, approach, and learning are the three main functions of reward. A rewarding stimulus is one that can induce the organism to approach it and choose to consume it. In neuropsychology, the *incentive salience model* distinguishes two components of an intrinsically rewarding stimulus (such as food), where "wanting" corresponds to appetitive (or approach) behavior and "liking" corresponds to consummatory behavior. In most cases, wanting and liking are superimposed. They are dissociated in certain cases. For instance, with cocaine addicts, wanting becomes separated from liking over time, as the desire to use the drug increases, while the pleasure obtained from consuming it decreases over time because of drug tolerance. This suggests that distinct "pleasure centers" exist within the brain to regulate *wanting* versus *liking*. Dopamine neurotransmission within the VTA system is thought to regulate the motivational aspect of pleasure (or *wanting*) and the pleasure centers outside the dopamine system (such as the ventral pallidum and parabrachial nucleus) are responsible for consummatory aspect of it (meaning the *liking* part). Animals in which the VTA dopamine system has been rendered inactive do not seek food, and will starve to death if left to themselves, but if food is placed in their mouths they will consume it and show expressions indicative of pleasure. Otherwise stated, they still like the food but are no longer motivated to work in order to acquire it. Hence, the activation of the VTA dopamine system makes the lover feel happy, alert, focused, and motivated to work in order to gain the affection of his or her loved one.

In a recent scientific study, the activity in the brains of seventeen subjects who were deeply in love were scanned while they viewed pictures of their partners, and compared with the activity produced by viewing pictures of three friends of similar age, sex, and duration of friendship as their partners (Bartels and Zeki, 2000). When viewing the pictures of loved ones, the brain activity was restricted to foci in the medial insula and the anterior cingulate cortex and, subcortically, in the caudate nucleus and the putamen which, together, form the aforementioned corpus striatum. Viewing photos of friends activated a different set of brain areas. As described previously, the corpus striatum is mainly in charge of the motor aspects of action selection. The insulae are believed to be involved in consciousness and play a role in diverse functions usually linked to emotion or the regulation of the body's homeostasis. These functions include compassion, empathy, perception, motor control, self-awareness, cognitive functioning, and interpersonal experience. The anterior cingulate cortex appears to play a role in a wide variety of autonomic functions, such as regulating blood pressure and heart rate. It is also involved in certain higher-level functions, such as attention allocation, reward anticipation, decision making, ethics and morality, impulse control, and emotion. The authors of the above study notice that studies of cocaine-induced euphoria have shown increased activity in foci that seem to overlap with all foci activated in their study—namely, the anterior cingulate cortex, the medial insula, the caudate nucleus, and the putamen, which suggests a potentially close neural link between romantic love and drug-induced euphoric states. It should not be a surprise at this point that the very same areas are involved in psychopathology, particularly in psychotic spectrum disorders.

Areas of the brain which are deactivated by love can also play an important role in the way we feel and act while we are in love. The widespread deactivations the authors of the above study have observed have their counterpart in previous studies which have shown that happiness correlates with deactivations in the right prefrontal and bilateral parietal and temporal

96

cortices. Sadness and depression correlate with activation in some of the cortical regions deactivated in their study, especially the right prefrontal cortex, of which the artificial inactivation by means of transcranial magnetic stimulation has proven to result in successful treatment against depression. The observed deactivation of the amygdaloid region is of special interest, since activity in it correlates with fear, sadness and aggression, and is thought to mediate emotional learning. Activity in this region increases from the most happy to the most fearful facial expression viewed. Their results show that, within experienced positive emotions, the amygdaloid region is more active when viewing friends than the loved partner. In a way, this should not come as a surprise. Many women report a visceral fear of pregnancy. It would only make sense that nature would try to subdue that fear so they may let themselves be impregnated. Notice that fear is mainly controlled by the amygdalae, the expression of which is reduced in the brain in love, while the activity in the locus coeruleus, which controls panic response and separation anxiety, is dialed up, at least in the attraction phase.

To sum up, the VTA-nucleus accumbens shell projection assigns incentive salience (or wanting) to cues associated with rewarding stimuli (such as the sight of the loved one), the VTA-orbitofrontal cortex projection updates the value of different goals in accordance with their incentive salience (for instance, by assigning top priority to conquering our beloved's heart), the VTA-amygdala and VTA-hippocampus projections mediate the consolidation of reward-related memories (marking memories of good times with our loved one for particularly easy retrieval), and both the VTA-nucleus accumbens core and substantia nigra-dorsal striatum pathways are involved in learning motor responses that facilitate the acquisition of rewarding stimuli (such as engaging in flirtatious behavior with the object of our passion to gain their affection).

Large amounts of dopamine are thought to induce psychosis. Indeed, many antipsychotic drugs target the dopamine system, have a broadly suppressive impact on most types of active behavior, and particularly reduce the delusional and agitated behavior characteristic of overt psychosis. Psychotic symptoms are often intensified by dopamine-enhancing stimulants such as methamphetamine or cocaine, and these drugs can also produce psychosis in healthy people if taken in large enough doses. Being in love causes the production of increasingly higher amounts of dopamine over time, leading the lover to experience a mental state not dissimilar from that of psychotic patients, especially those who are also manic.

Serotonin

Serotonin, or 5-hydroxytryptamine (5-HT), is a monoamine neurotransmitter. It is primarily found in the enteric nervous system located in the gastrointestinal tract, where it regulates intestinal movements. It is also produced in the central nervous system, specifically in the raphe nuclei located in the brain stem. Additionally, serotonin is stored in blood platelets and is released during agitation and vasoconstriction. It is a growth factor for some types of cells, which may give it a role in wound healing. In the central nervous system, serotonin is involved in regulating mood, appetite, and sleep. Notable for our purposes is the fact that elevated mood, decreased appetite, reduced gastrointestinal activity, and decreased need for sleep are noted in most studies of people in love. Serotonin is also involved in cognitive functions such as memory and learning. Additionally, it plays a role in the psychedelic effects of recreational drugs such as hallucinogens. While production of dopamine and norepinephrine are positively correlated to each other, that of serotonin inversely correlates with both, meaning high levels of dopamine and norepinephrine (detected, for instance, in the brain of someone in love) typically imply low levels of serotonin.

Recent findings suggest that the serotonin transporter might be linked not only to neuroticism and sexual behavior, but also to obsessive-compulsive disorder (OCD). High levels of neuroticism correlate positively with psychopathological states such as anxiety and depression. Obsessional thinking in the form of rumination is very typical of depressed patients. In fact, many antidepressants work by raising the level of serotonin by inhibiting excessive reuptake by serotonergic neurons. Manic people often become obsessed with new pursuits. Many among those diagnosed with OCD perform compulsive rituals (such as repeatedly washing their hands) because they inexplicably feel they have to, while others act compulsively so as to mitigate the anxiety that stems from particular obsessive thoughts. Obsessively thinking about the loved one, compulsive flashbacks of joyful moments with them, and sexual longing are common in individuals in love. The similarities between obsession and an overvalued idea—such as that typical of subjects in the early phase of a love relationship—prompted a few scientists to explore the possibility that the two conditions might share alterations at the level of the serotonin transporter (Marazziti et al., 1999). They recruited twenty individuals who had fallen in love within the previous six months, twenty unmedicated OCD patients, and twenty baseline controls. The main finding of the study was that subjects who were in the early romantic phase of a love relationship were not different from OCD patients in terms of the density of the platelet serotonin transporter, which proved to be significantly lower than in the normal controls.

Testosterone

Testosterone is a sex hormone produced by the testes in men, to a far lesser degree by the ovaries in women, and by the adrenal glands in both men and women. It is involved in libido, vitality, aggression, bone density, fat distribution, muscle strength and mass, face and body hair, the production of sperm in males, and the production of red blood cells in both men and women. Low testosterone can lead to decreased sexual desire, changes in sleep, reduced muscle mass and bone density, and decreased self-esteem. The hormone is also thought to suppress anxiety and have a protective effect against depression. The production of testosterone is controlled via two specific feedback loops called the hypothalamic-pituitary-gonadal axis and the hypothalamic-pituitary-adrenal axis. The hypothalamus secretes gonadotropin-releasing hormone (GnRH), which signals the pituitary gland to secrete luteinizing hormone (LH), which in turn signals the gonads and the adrenal glands to produce testosterone.

The relationship between the neurotransmitters dopamine, norepinephrine, and serotonin on the one hand, and testosterone on the other hand, is complex. High levels of dopamine are correlated with high levels of norepinephrine and are also found to induce high levels of testosterone, and vice versa. While high levels of dopamine are correlated with low levels of serotonin, higher levels of testosterone are found to lead to higher levels of serotonin, meaning testosterone may suppress some of the effect of dopamine on serotonin levels. To complicate matters more, testosterone also modulates the effects of the bonding hormones, oxytocin and vasopressin. It enhances the effect of vasopressin and reduces that of oxytocin. Women, on average, tend to have lower levels of testosterone and lower libido than men and are also more prone to both depression and anxiety than men. As they grow older and as their sex hormone output falls, men suffer more commonly from depression and anxiety, and some studies have already shown a positive effect of testosterone supplementation on the moods of the test subjects. A recent study demonstrated that testosterone supplementation increases the number of serotonin transporters in the human brain (Kranz et al., 2014). Many antidepressants also work by increasing brain levels of serotonin.

Being in love appears to lower testosterone levels in men while increasing it in women (Marazziti and Canale, 2004). Could the surge in the bonding hormone vasopressin depress testosterone in men? One may presume that these changes in testosterone result in the temporary reduction of differences in behavior between the sexes. While women may seek sexual activity more assertively, men may become more interested in emotional bonding. Men who produce less testosterone are more likely to be in a relationship or married, and men who produce more testosterone are more likely to both engage in extra-marital relationships and to divorce. They tend to show more aggressive and competitive behaviors. Testosterone is responsible for the masculinization of the fetal brain. Individuals, whether male or female, with masculinized brains due to prenatal and circulating androgens are thought to use aggression as a way of enhancing their resource acquiring abilities in order to survive and to attract and copulate with as many mates as possible. Fatherhood is found to decrease testosterone levels in men, suggesting that the emotions and behaviors tied to decreased testosterone may promote paternal care. Postpartum depression is sometimes observed in new fathers, especially when their spouse is also depressed postpartum. Decreased testosterone levels may partly be involved in triggering the condition.

Oxytocin, Vasopressin, and Endorphins

Oxytocin and vasopressin are two biochemicals with similar structures and are majorly involved in pair-bonding. They are coded by two genes that are located close to each other on the same chromosome. The two genes are believed to have resulted from a gene duplication event which occurred some 500 million years ago. Nerve cells in the hypothalamus make and transport them to the pituitary gland, which then releases the hormones into the bloodstream. They are produced in both men and women, albeit in differing amounts, and may differentially affect them. The same brain areas also produce powerful biochemicals called endorphins named *alpha* (α), *beta* (β), *gamma* (γ), and *sigma* (σ) endorphins. Endorphins are morphine-like substances produced within the body and endowed with the same painkilling and euphoric effects as synthetic opioids. They attach to our cells' opioid receptors just like medicinal painkillers or recreational opioids do. Endorphins are released when a person gets hurt, but also during exercise, laughter, or sex. The common denominator between oxytocin, vasopressin, and endorphins is not just their coproduction in the same brain areas, but also their role in facilitating human social and sexual bonding.

In the body, oxytocin is thought to modulate inflammation. Thus, the increased release in oxytocin following positive social interactions has the potential to improve the healing of wounds. This may be what is behind the adage that having strong affiliative relationships confers health and longevity benefits. Oxytocin also modulates genital contractions during both female and male orgasm, uterine contractions during delivery, and muscular contractions during lactation. The flood of oxytocin in the last moments of delivery (and that of endorphins throughout labor) is what makes women forget the pain of pregnancy and labor. It is similar to being hit with an extreme dose of heroin. The women for whom this process is impaired may be prime candidates to develop postpartum PTSD, in which they are haunted by frightening flashbacks of painful moments during labor. The same women may fail to properly bond with their baby and suffer from postpartum depression, given that oxytocin plays a very important role in initiating maternal behavior and parent-offspring bonding. Female rats given oxytocin-antagonists after giving birth do not exhibit typical maternal behavior (Van Leengoed et al., 1987). By contrast, virgin female rats show full maternal behavior toward foreign pups upon administration of oxytocin directly into their brains (Pedersen and Prange, 1979). For this last study, criteria in five behavioral categories had to be fulfilled by an animal within two hours of injection for its behavior to be considered fully maternal. The authors also found that the presence of estrogen may be necessary for the induction of full maternal behavior by oxytocin. In contrast, none of the saline- or vasopressin-treated animals displayed full maternal behavior. Oxytocin is also associated with feelings of pity and the willingness to help vulnerable individuals. Testosterone, which men produce in copious amounts, may act to inhibit oxytocin's effect to some degree, which is thought to have been useful ancestrally since pitying the animals one was about to kill would not have been a particularly valuable instinct for a hunter. Women produce far more oxytocin than men and have less testosterone to suppress its effect. Is it any surprise, then, to see more women choosing social work education in college, going into care-related professions, or becoming the unpaid caregiver of an aging parent? In romantic relationships, oxytocin evokes feelings of contentment, reductions in anxiety, and feelings of calmness, trust, and security when in the company of one's mate. This suggests that oxytocin may be important for the inhibition of the brain regions associated with behavioral control, fear, and anxiety. Thus, it has the ability to majorly impair our decision-making ability by making us naive and trusting. Otherwise stated, while in love, our judgement may well be toast; love, truly, makes us blind. In a social context, oxytocin can increase positive attitudes toward individuals with similar characteristics who then become classified as "in-group" members, whereas individuals who are dissimilar become classified as "out-group" members. Oxytocin has been associated with lying when doing so would benefit in-group members. It is thought to be a driving force behind both ethnocentrism and group conformity.

Women are generally found to be more conformist than men. Could their higher oxytocin levels account for some of it? The difference between high and low levels of parent-offspring bonding, romantic bonding in monogamous relationships, bonding in other kinship-based relationships, bonding in friendship, and that between compatriots may simply be a matter of degree or biochemical dosage from the highest levels of oxytocin to progressively lower levels. Dysfunction of the oxytocin system in the form of oxytocin hyposensitivity or deficiency is thought to be involved in autism (mostly affecting men) which is characterized by strong social deficiencies. A different type of dysfunction of the oxytocin system in the form of oxytocin hypersensitivity or overproduction may be behind one atypical depression in which affected persons (who are mainly women) have a continually depressed baseline from which they can spring out in an instant and be rapidly cheered up by falling in love or by forming other strong social bonds. They have a propensity to attach themselves to people very quickly and display dependent and clingy behaviors in relationships—even unhealthy ones. Many battered women would fit this profile.

Both men and women naturally produce vasopressin, yet men experience its effects more strongly because of how it interacts with the male sex hormone testosterone. For men, vasopressin plays a role similar to that of oxytocin in women when it comes to forming monogamous pair-bonds. A species of bird called voles comes in two flavors: the highly monogamous prairie voles and the highly promiscuous montane voles. When otherwise promiscuous montane voles are dosed with oxytocin and vasopressin, they adopt the monogamous behavior of their prairie cousins. Once a male and a female prairie vole fall in love at the beginning of the mating season, they feel an urge to have sex together with a mind-boggling frequency. In doing so, they produce copious amounts of oxytocin and vasopressin. Following their marathon sex sessions, the male starts to build a nest and guard its mate, much like a good husband. It turns out that the difference between the males of the montane versus the prairie kind is in the length of the gene that codes for vasopressin; it is longer in the latter than in the former. Human males also differ in the length of the same gene: the men with the longer versions are more monogamous and those with the shorter versions tend to be more promiscuous. The presence of testosterone may help vasopressin to accomplish its mission, while that of estrogen may, to some degree, inhibit it, producing a gendered effect. It follows that a genetic test could majorly help women desirous of a lifelong relationship in selecting the right long-term partner. Vasopressin has other functions in the body, such as acting as an antidiuretic and a vasoconstrictor.

As mentioned previously, the main difference between the endorphins (our body's natural opioids) and the analgesic drugs (or synthetic opioids) is that the former are cleared from the blood far more quickly than the latter. Endorphins are also involved in the release of sex hormones from the brain's pituitary gland. It is no surprise that they should be a part of the love cocktail of biochemicals: they make us feel good and rev up our desire to be sexual. Release of endorphins has a strong positive correlation with that of oxytocin, vasopressin, dopamine, norepinephrine, and testosterone. The term "endorphin rush" (meaning the feeling of being high or full of energy and vitality) is sometimes used in popular parlance to refer to a feeling of wellness caused by exercise, danger, or light to moderate stress. By the same token, endorphins may be involved in the sensation of rush we feel when we fall in love. Love causes the mental representation of our loved one to become permanently associated with pleasure in our brain. Constantly recalling memories of pleasurable moments with them, just like repeated actual contact, leads to repeated release of these powerfully addictive biochemicals. Over time, we become hooked to our loved one the same way that an addict becomes hooked to heroin. The dual effect of oxytocin and endorphins released during lactation may lead to a similar sort of dependency between mother and baby. Many otherwise ambitious professional women report a loss of interest in returning to work following a parental leave of absence. Many who return early report feeling separation anxiety or pain, resembling symptoms of sudden opioid withdrawal. Because they act to alleviate pain, endorphins may be released during acupuncture and massage as well, causing us to feel good and relaxed at the same time.

Keith Kendrick provides a good summary of the above discussion (Kendrick, 2004): "When released in the brain through giving birth or mating, the neuropeptides oxytocin and vasopressin are involved in promoting parent-offspring and monogamous bonds in animals such as sheep and voles. . . . There is no point in being able to bond with another individual unless you can recognize them. Brain neuropeptides like oxytocin and vasopressin, which are known to be associated with formation of social bonds, are also involved in promoting social recognition memory. . . . In females that bond selectively with offspring after giving birth, or form monogamous pair-bonds after sex, oxytocin receptors are highly expressed in a region of the brain called the nucleus accumbens and the peptide can facilitate brain dopamine release. . . . The same is true of the vasopressin system which promotes bonding responses in males, although the ventral pallidum is the critical dopamine-producing reward site involved. . . . There is a strong link between oxytocin and vasopressin systems in the brain and modulation of the release of brain opioid peptides. . . . In both monkeys and humans, turning down the gain in opioid reward centers in the brain by blockade of μ-receptors with drugs will actually induce individuals to seek social contact, and this has been used as a therapeutic approach in autism. . . . Indeed, a seeming paradox is that the neurochemical systems involved in forming and maintaining social bonds are also potent stimulators of anxiety. It appears that social species have evolved a dual mechanism for ensuring that social bonds are both sought and maintained—you are anxious until social contact is achieved and once this happens your anxiety is hopefully, although not necessarily, replaced by feelings of pleasure. . . . The major role of these peptides may simply be the formation of social bonds and linking social recognition systems in the brain with those that make us feel pleasure. Thereafter, perhaps continued episodes of release merely act to reinforce or at least help maintain these links for long periods. . . . Brain imaging studies on individuals viewing pictures of their romantic partners or newborn babies have confirmed that oxytocin- and vasopressin-containing regions and dopamine reward centers are indeed particularly involved. So perhaps love really is a simple matter of chemistry and animal attraction after all!"

Nerve Growth Factor (NGF)

The nerve growth factor (NGF) is a neuropeptide involved primarily in the growth, maintenance, proliferation, and survival of nerve cells (or neurons). In fact, NGF is critical for the survival and maintenance of sympathetic and sensory neurons, that of pancreatic beta cells, and the regulation of the immune system. NGF prevents or reduces neuronal degeneration in animal models of neurodegenerative diseases and promotes peripheral nerve regeneration and myelin repair. Myelin is the fatty tissue that insulates the axons of nerve cells to increase the speed at which information (encoded as an electrical signal) travels from one nerve cell body to another (very much like the coating of copper wires used to transmit electricity). Dysfunction of the NGF signaling may be involved in various psychiatric disorders, such as depression, bipolar disorder, schizophrenia, autism, anorexia, and bulimia nervosa, as well as neurodegenerative conditions such as multiple sclerosis (MS), amyotrophic lateral sclerosis (ALS; also called Lou Gehrig's disease), dementia (including Alzheimer's disease), Parkinson's disease, and Rett syndrome (which is caused by a de novo mutation on the X-chromosome and primarily affects girls; its symptoms include problems with language, coordination, repetitive movements, slower growth, problems walking, smaller head size, sleeping problems, seizures, and scoliosis).

When it comes to romantic love, a recent study found that the concentration of NGF in the blood plasma was significantly higher in individuals who had been in a romantic relationship with another person for less than twelve months, than those who were either not in a romantic relationship or had been in one for more than twelve months (Emanuele et al., 2006). In the thirty-nine subjects in love who—after twelve to twenty-four months—maintained the same relationship but were no longer in the same mental state to which they had referred during the initial evaluation, plasma NGF levels decreased and became indistinguishable from those of the control groups. NGF could be critical in the formation of the links between the mental representation of our loved one and our brain's pleasure centers. It could, in conjunction with the hormones and neurotransmitters we previously reviewed, help to mark the memories of pleasurable moments with our loved one, along with the associated positive emotions, for easy retrieval. The study is also strong proof that we do not stay in this state forever. The average duration may be about a year, and the differences among individuals could be vast with duration varying from a few months to two or three years. An evaluation of the range of duration would require a much larger scale (and more costly) study.

Cortisol and Epinephrine

Cortisol is a steroid hormone released by the adrenal cortex, while epinephrine (or adrenaline) is a monoamine produced in the adrenal medulla. The production of both hormones is controlled via the hypothalamic-pituitary-adrenal axis feedback loop. The hypothalamus secretes gonadotropin-releasing hormone (GnRH), which signals the pituitary gland to secrete luteinizing hormone (LH), which in turn signals the adrenal glands to produce cortisol and epinephrine. They are released with a diurnal cycle (with levels peaking early in the morning and bottoming out at night) and their release is increased in response to stress and low blood glucose concentration. Cortisol and epinephrine levels are also elevated when we fall in love. In the body, both hormones play a role that is similar in many ways to that of norepinephrine during the fight-or-flight response and can have physiologically deleterious consequences in cases of prolonged exposure to high levels of stress.

When it comes to romantic love, the combined action of cortisol and epinephrine is twofold: the hormones may contribute to the feeling of rush (meaning the sensation of vitality, zest, and vigor) we experience when we fall in love and play a major role in the formation of flashbulb memories (the vivid memories of pleasurable moments with our loved one that we keep replaying in our heads). The contribution to the feeling of rush is directly tied to the action of these hormones on our metabolism, which consists of increasing blood glucose levels for conversion into energy. Epinephrine also promotes alertness. In addition, cortisol works with epinephrine to create memories of short-term emotional events; this is the proposed mechanism for storage of flashbulb memories, and may have originated as a means to remember what to avoid in the future (in the case of particularly unpleasant situations) and what to vie for in the future (in the case of particularly pleasant ones). Hence, PTSD (or post-traumatic stress disorder) and falling in love share the same process for the creation of memories involved in the compulsive flashbacks that the affected persons experience.

A flashbulb memory is a highly detailed and exceptionally vivid mental snapshot of the moment and circumstances in which an event of a surprising and consequential or emotionally arousing nature occurred. A number of studies have found that flashbulb memories are formed immediately after a life-changing event happens or when news of the event is relayed. The intensity of the emotional state of the individual directly contributes to the creation of a flashbulb memory. To strengthen the association, thus enabling the individual to vividly remember the event, emotional state and affective attitude contribute to overt mental rehearsal of the event to strengthen the memory of the original event which, in turn, determines the formation of a flashbulb memory (Finkenauer et al., 1998). Our brains tend to have a hard time telling the difference between what is real and what is not. As the lover mentally replays a pleasurable moment spent with their loved one, his or her body releases the biochemical cocktail made up of all the hormones and neurotransmitters we just reviewed, causing him or her to feel the original emotions associated with those memories as if the event was really happening again. Mental rehearsal not only strengthens the link between the mental representation of the loved one and the sensation of pleasure, but it also creates dependency (or attachment) resembling addiction to recreational drugs. Repeated enough times, a stranger about whom we may truly know very little, ends up becoming a familiar figure in our life alongside our kin.

6

Interplay of Psychopathology and Love

I have thus far mainly focused on the relatively healthy course of passionate love (referring to the core of what we call "falling in love" or stage two in my psychodynamic model) and its aftermath if pair-bonding is successful, or what is usually termed "companionate love" (referring to stage four and beyond in my model). Rejection in the second stage of my model has two consequences in the normal, nonpsychopathological case, where the lover accepts it: Firstly, stage three occurs sooner, cutting short the duration of the second (psychotic amplification) phase. Here, a prefrontal cortex with a good or better level of cognitive control over the limbic system shuts down the flow of the biochemical cocktail that triggered the state of being in love rather quickly. Secondly, stage four of my model still takes place without any modification to its duration of four to five years. What remains after all is said and done is a fond memory of the person we fell in love with. A breakup in stage four or later, when it proceeds with calm and control, results in psychological pain which dissolves itself reasonably fast as the two former partners each move on their own path following legal, financial, or other practical negotiations.

Preexisting psychopathology and falling in love, however, do not always mesh well together. Unrequited love is love that is not reciprocated and can sometimes have severe consequences when the lover refuses to accept defeat and move on, including a host of negative emotions and destructive behaviors such as anger, jealousy, desire for revenge, low self-esteem, rumination, hopelessness, feelings of loneliness and anguish, clinical depression, stalking, violence, and in the most unfortunate cases, suicides or homicides. I mentioned in chapter 3 how we seem to have an intuitive and acute knowledge of our own mate value and tend to seek individuals whom we perceive to be of roughly equal mate value to maximize our chances of reproductive success. Many mental disorders can distort the way that we perceive ourselves, leading us to either overestimate or underestimate our mate value. The hypomanic or manic phase of bipolar disorder, for instance, often leads to a hyperbolic sense of self, which can make us more prone to fall in love with someone of much higher mate value than ourselves, meaning someone who is generally judged to be more attractive because of their superior physical beauty, intelligence, socioeconomic status, or other attributes. Love born under such conditions is inherently more prone to become unrequited. These conditions also make people more prone to both obsessions and socially aggressive and persistent behaviors, which can lead to stalking in their extreme manifestations. If rejection is felt to be overly stressful, the bipolar individual's mood can rapidly flip in the opposite direction, leading to feelings of worthlessness, low self-esteem, and depression, which can include suicidal ideation and behaviors. Dysthymia, or persistent depressive disorder, can lead to excessive and persistent self-deprecation, low self-esteem, and an overall feeling of inadequacy. Within the context of an existing intimate relationship, when one partner's sense of their own mate value is particularly low, he or she may engage in emotional or physical abuse when threatened with abandonment in their desperation to hang on to their current mate. When scientists imaged the brains of ten women and five men who had recently been rejected by a partner but reported that they were still intensely, albeit unhappily, in love (Fisher et al., 2010), they saw activations in brain areas associated with gains and losses, craving, and emotion regulation, which were, in fact, some of the same areas activated when we are happily in love, using recreational drugs, or suffering from an episode of mental illness—those that I previously characterized as being disorders of our brain's attraction or attachment systems.

Additionally, the falling-in-love process we have evolved includes elements which make it intrinsically liable to mismatching, including its timing (coinciding with moments of intense emotional arousal), the sketchiness of our ancestral checklists for attributes we want in a reproductive partner (which entirely omit the fluid nature of sex, gender, sexual orientation, or interest in parenting), and the psychotic nature of passionate love (which makes us detached from reality and impairs our judgment). The object of unrequited love is often a friend or acquaintance, someone we regularly encounter in the workplace, school, or other group activities in which we may be engaging. To illustrate, here is a description of a specific case of unrequited love by an anonymous participant to an online chat room: "E is my best friend and I'm his best friend. . . . Our friendship is honestly the best thing I have. . . . I am also in love with him. It hurts like hell because he's literally the cutest, nicest, most amazing guy I've ever met in my life. I honestly want to spend the rest of my life with him, and I care about him deeply, but I know for a fact he'll never feel the same way. I forgot to mention (that) I'm a male and E is also a male. He's very accepting when it comes to homosexuality but he's straight. That's the problem. I'm in love with my straight best friend. . . . I often find myself thinking about suicide. If I can never be with such a beautiful person as him, then I don't really see the point in living, because that's honestly the only thing I want in my life." Here is another anonymous posting: "This problem is engulfing my life right now and causing me enormous amounts of anxiety and intense sadness. . . . I'm in love with one of my best friends. I'm 99% sure she is not at all interested in being more than friends. This is a very common problem, but (knowing) that doesn't make the pain less real. It is a roller-coaster ride where spending time with her is euphoric, time apart is painful, and hearing about her (being romantic) with other guys is soul crushing. . . . So here I am in constant pain with no one to talk to and nothing to do about it." Here is a third posting along the same lines: "Unrequited love has got to be the most painful and agonizing thing that a person can encounter. I have never come across anything as devastating in my entire life. . . . I feel that my chances of finding love were flushed down the toilet as soon as this person got married

and moved on with somebody else. . . . I'm caught between two decisions: 1.
Kill myself and put an end to all of my misery. The thought of this person
being around somewhere and not being with them is like hell on earth or;
2. Live out the rest of my life in misery and fight on. All of our lives go by
pretty quickly anyway, don't they?"

In extreme cases, lovesickness does drive young people to take their own
lives. A recent analysis of the suicide notes left in forty cases of suicide in
India (Bhatia et al., 2006) revealed that a disturbed love affair was the most
common reason mentioned followed by financial problems. In a majority
of notes, there were indications of hopelessness and depression (52.5%). A
majority of suicide note writers were twenty-one to thirty years of age (55%)
and were males (65%). Women are generally known to attempt suicide more
often than men, but men have a much higher rate of suicide completion
than women. A larger study conducted in Australia (Lester et al., 2004)
analyzed 262 suicide notes and found that women were significantly more
likely to have "escape from unbearable pain" as a motive in their suicides but,
contrary to myth, were significantly less likely to have love and romantic
problems as a precipitant. It is a general finding that men seem to experience
the most distress in the early stages of romantic love (the attraction phase),
while women seem to get mostly caught up emotionally in the later stages
(the attachment phase). This may have an evolutionary basis: men have had
most to gain in attracting a mate, impregnating her, and then moving on
to another mate, while women have had most to gain in keeping a mate
around for the long run to ensure continuous provisioning and protection
for themselves and their offspring. According to Kay Redfield Jamison, the
overwhelming majority of adolescents and adults who commit suicide have
been determined, through postmortem investigations, to have suffered from
bipolar manic-depressive or unipolar depressive illness. The hormonal roller
coaster of unrequited love may, in extreme cases, cause individuals with a
fragile biochemical makeup to commit the ultimate act of self-destruction.

On an equally sinister note, if feelings of hopelessness and sadness can
lead to suicide, those of jealousy and rage can trigger homicides. A "crime
of passion" refers to a homicide in which the perpetrator commits the act

against someone in the "heat of passion" or out of sudden rage rather than as a premeditated crime. Until recently, the offense was forgivable in most countries. Even in today's France, *le crime passionnel* only carries a light prison sentence. In the US, a successful "crime of passion" defense may result in a conviction for manslaughter or second-degree murder instead of first-degree murder, because a conviction for the latter requires premeditation. Many defendants will opt for "temporary insanity" or "provocation" defenses to extricate themselves from the charges. Most perpetrators are male and most victims are female. According to a recent report released by the Centers for Disease Control and Prevention (Petrosky et al., 2017), over half of the killings of American women (55.3%) are related to intimate partner violence, with the vast majority of the victims dying at the hands of a current or former romantic partner. Approximately 15% of female victims of reproductive age (eighteen to forty-four years old) were pregnant or less than six weeks postpartum. Over 11% of victims of intimate-partner-violence-related homicides experienced some form of violence in the month preceding their deaths, and argument and jealousy were common precipitating circumstances. Findings indicate that young women, particularly racial/ethnic minority women, were disproportionately affected. The sentiment of jealousy has been evolved in the context of mate guarding, primarily by males in their attempt to ensure paternity certainty. Of course, women are prone to the feeling of jealousy as well, but their mate-guarding efforts tend to take more passive-aggressive forms, a likely consequence of the historically bigger and sturdier male physicality. Probably for the same reason, the female brain has evolved stronger inhibition mechanisms for anger than the male brain. Women are also more likely to direct their aggression towards themselves (as in self-mutilation acts, such as cutting oneself) or towards the woman with whom they suspect their intimate partner to be having an affair. Another gendered aspect of jealousy is that men tend to find sexual infidelity most distressing, while women tend be most disturbed in cases of emotional infidelity. The divergence has, once again, been shaped by evolutionary pressures and has its roots in the notion of differential parental investment: men's distress is related to ensuring

113

paternity certainty (by guarding their partner's biological resources) while women's distress is related to ensuring provisioning certainty (by guarding their partner's material resources). The men endowed with the dark triad traits of narcissism, Machiavellianism, and sociopathy may be more likely to commit homicide and to engage in violent behavior within their relationship. Research shows that these traits are more frequently found in men than in women. This is not surprising, since preliterate societies are known to breed such men: those who have killed are awarded with social status, resources, and fertile women to impregnate. Modern practice consists in locking them up in prison cells. In these cases, although some form of violent act has already been committed, it may hamper further social harm and, perhaps more importantly, further propagation of their traits.

On the issue of violence within male-female relationships, research on marital violence and dating violence is increasingly merging: at least some of the couples displaying violence during courtship are now believed to eventually marry and continue a pattern of violence (Frieze and Davis, 2000). As for the issue of stalking, the most common form of stalking tends to occur during the breakup phase of relationships, although it may also occur during initial courtship before a true relationship has begun. Stalking is, in fact, often most violent during breakups, and is perpetrated most by those who had been physically abusive during the relationship. Both sexes are found to be engaging in this type of assaultive behavior. Women are often found to engage in more of the low-level types of violence that characterize most physically abusive relationships. Defining stalking is tricky, and many feel they are being stalked even though they may not meet current legal definitions. The website for the Stalking Resource Center proposes the following working definition of stalking: it is "a course of conduct directed at a specific person that would cause a reasonable person to feel fear." It lists the following among the things a stalker may do: follow you; send you unwanted gifts, letters, cards, or e-mails; damage your home, car, or other property; monitor your phone calls or computer use; use technology to track you; drive by or hang out at your home, school, or work; threaten to hurt you, your family, friends, or pets; find out about you by using public records or online search services; hiring investigators; going through your garbage; contacting friends, family, neighbors, or co-workers; posting information or spreading rumors about you on the internet, in a public place, or by word of mouth; undertaking other actions that control, track, or frighten you. A publication by the Centers for Disease Control and Prevention (Breiding et al., 2014) reports the following statistics with regard to sexual violence, stalking, and intimate partner violence in the twelve-month period between January and December 2011: in the United States, an estimated 19.3% of women and 1.7% of men have been raped during their lifetimes; an estimated 1.6% of women reported that they were raped in the twelve months preceding the survey; an estimated 43.9% of women and 23.4% of men experienced other forms of sexual violence during their lifetimes, including being made to penetrate, sexual coercion, unwanted sexual contact, and noncontact unwanted sexual experiences (the percentages of women and men who experienced these other forms of sexual violence victimization in the twelve months preceding the survey were an estimated 5.5% and 5.1%). An estimated 15.2% of women and 5.7% of men have been a victim of stalking during their lifetimes, and an estimated 4.2% of women and 2.1% of men were stalked in the twelve months preceding the survey.

Having a mental disorder (just like having a physical or intellectual disability) generally raises an individual's prospects of becoming either a perpetrator or a victim within an abusive relationship. Codependency is a behavioral condition in a relationship where one person enables another person's addiction, poor mental health, immaturity, irresponsibility, or underachievement. On average, more men fit the profile of the abuser. They use tactics such as psychological abuse (in the form of condescension and contempt) in order to lower their partner's self-esteem and diminish their perceived value on the mating market if they are contemplating a breakup, or coercion escalating into physical violence as their attempt at controlling their partner's sexuality results in increasingly higher frustration. Meanwhile, they continue providing material resources to their mates to keep them around. On average, more women tend to fit the profile of the victim. Many display tendencies observed in people with dependent personality disorder or fit the profile of those colloquially described as "relationship junkies" or "applause junkies." The latter have a condition which is clinically termed "atypical depression with rejection sensitivity." The baseline of their psychological makeup is characterized by depression and anxiety. They can, however, be lifted up from this baseline and cheered up in an instant by falling in love, rapidly bonding with others, or seeing others applaud them. Approval from others is paramount to them, more so in many cases than self-respect. These people move toward others by gaining their approval and affection, and subconsciously control them through their dependent style—which, in a way, is both assaultive and passive-aggressive. They are seemingly unselfish, virtuous, martyr-like, faithful, and turn the other cheek despite personal humiliation. They tend to emotionally overwhelm their partner with demands for love and commitment reassurances. The slightest sign of commitment slip from their partner will send them into a panic. Many are not very choosy when it comes to selecting a partner, as long as the latter does not demand much in the way of independence from them, making them easy prey for men endowed with the dark triad traits. These women tend to cling to their partners despite some of the worst abuse, displaying many behaviors typical of stalkers. As soon as a relationship ends, they sink into a clinical depression, from which they try to emerge by making desperate attempts to immediately replace their ex-partner. They feel some of the most extreme forms of loneliness and anguish when they are not in a relationship. Since they are not very choosy, they end up meeting another mate whose character is similar to the one they just lost. They fall in love with them instantly. Some are ready for a marriage proposal within days of the first date. They attach quickly and strongly, in what seems like a disorder of the brain's attachment circuitry. They cling. And the cycle starts over. Another typical long-term pairing is one between a woman suffering from borderline personality disorder (a condition which has a lot of commonality with the previous disorder) and a man with narcissistic, Machiavellian, or antisocial personality disorder. These relationships are some of the most turbulent and, oftentimes, violent ones as the couples always seem to be on the verge of a breakup—but these relationships are, in reality, some of the most enduring kind.

Besides the disorders I mentioned above, a number of specific mental conditions have been identified which are directly associated with the subject of love. Mental health providers disagree, however, about whether any of them should be classified as a stand-alone disorder or simply as a symptom of anxiety and psychosis spectrum disorders. Perhaps they are symptoms of the latter, which only appear comorbidly when the affected individual is also in love. What follows is a partial list of these disorders.

Erotomania

Erotomania is a type of delusional disorder which mainly affects women of a shy and dependent temperament, who also tend to be sexually inexperienced. The delusion consists of the unshakeable belief that an unattainable man (given his socioeconomic standing, marriage, or simple disinterest) is secretly infatuated with them. The object of their passion can be a casual acquaintance (e.g., the driver of a Maserati whom they regularly see at the local pastry shop around breakfast time), a professional service provider (e.g., their primary physician, dentist, or accountant), a member of the senior management team at work, or a celebrity (e.g., a movie star or rock star whom they have never even met in person). A concurrent delusion they all have is one in which their perceived admirer secretly communicates their love to them by subtle methods such as facial expressions, manner of speaking or gesturing, body posture, choice of words or clothes, a combination of words or expressions they may use, arrangement of furniture, colors, and other seemingly innocuous acts, or, if the person is a public figure, through clues in the media. The affected individuals often engage in behaviors characteristic of stalkers. Erotomania has two forms: primary and secondary. Primary erotomania is commonly referred to as *de Clerambault's syndrome* and *old maid's insanity*. It exists alone without comorbidities, has a sudden onset, tends to be irresponsive to treatment, and typically lasts a lifetime. The psychiatrist G. G. de Clerambault first described the disorder in a female patient who was obsessed with the British monarch George V. She would stand outside of Buckingham Palace for hours at a time, believing that the king was communicating his desire for her by moving the curtains. The secondary form has a more gradual onset and is found along with mental disorders like paranoid schizophrenia or bipolar disorder, which often include persecutory or grandiose delusions and hallucinations. For instance, a sufferer could, over repeated episodes of mania spanning many years, become persuaded that a handsome news anchor is using a secret code to communicate with her during daily newscasts. She starts to initiate contact with him by sending flattering letters addressed to him to the news station where he works. Then, she runs background checks on him which reveal his home address and cell phone number. She gets in the habit of passing by his house every day and randomly calling him on evenings. When he places a restraining order on her, she believes that this is a secret message encouraging her to pursue him even more forcefully. This secondary form of erotomania can usually be effectively treated with a combination of antipsychotics and psychotherapy. Of course, erotomania does impact men as well as women, albeit less frequently. For instance, John Hinckley, Jr., the man who attempted to assassinate President Ronald Reagan is reported to have had erotomanic delusions toward Jodie Foster, and may have attempted the assassination out of misguided desire to impress the actress.

Hypersexuality disorder

This disorder characterizes individuals who are often the target of many comedies, late-night talk shows, or standup comedian's jokes, and depicted as those attending sex addiction counseling meetings. All jokes set aside, it has the potential to destroy relationships, break apart families, and cause the spread of sexually transmitted diseases among other ills. People with hypersexuality disorder experience consistent but unsuccessful attempts to control or diminish the amount of time spent engaging in sexual fantasies, urges, and behaviors in response to dysphoric mood states (such as loneliness, depression, and anxiety) or stressful life events. Their condition is so severe as to cause distress, danger to themselves or others, or impairment in social, occupational, or other important life functions. It is estimated to affect 3 to 6 percent of adults in the United States, predominantly male. It was proposed for inclusion in *DSM-5* (the *Diagnostic and Statistical Manual of Mental Disorders, Fifth Edition*), but ultimately declined. The decision may, in part, be based on the fact that the condition often does not stand alone, but manifests itself among the multiple symptoms of other mental disorders, such as the manic phase of bipolar disorder or the hypomanic phase of borderline personality disorder. Tagging it as a disorder could also lead to wrongly applying it to individuals who are naturally endowed with a high libido, an above-average risk-taking temperament, and capable of perfectly sound judgement and self-control. The main differentiator is the fact that people with hypersexuality disorder report feeling out of control, and act on their sexual urges while disregarding the repercussions. Some critics also worry that hypersexuality disorder could be used as an excuse to be unfaithful.

Othello Syndrome or Delusional Jealousy

This is a psychotic condition which primarily affects men. If we remember the fact that men may be the ones who predominantly evolved the sentiment of jealousy in their ancestral obsession to ensure paternity certainty, then it should not come as a surprise that they would be the ones endowed with the most extreme versions of it. The name is derived from the Shakespearean play *Othello: the Moor of Venice*, in which the main character, Othello, becomes so convinced about the infidelity of his wife, Desdemona, that he murders her, and then commits suicide. Like Othello, there are people who develop the unshakeable and obsessional belief that their partner is being sexually unfaithful without any facts to support it. They tend to misinterpret the behavior of their partner to provide evidence for their distorted perception. They may engage in such conduct as isolating their partner from their family, friends, and employment and not letting the partner have personal interests or hobbies outside the house; constant surveillance; interrogating their partner about phone calls and all other forms of communication; going through the partner's belongings; accusing the partner of looking at or giving attention to other people; claiming the partner is having an affair when they withdraw from sexual activity; strict and detailed rules for behavior; restrictions on access to such basic necessities as food, clothing, and sanitary facilities; verbal or physical violence towards the partner or the individual who is perceived to be the rival; blaming the partner and establishing an excuse for jealous behavior; and threatening to harm others or themselves. Lucy Vincent notes that mate-guarding efforts such as "showcasing extreme vigilance," "mate sequestration," and "belittling" are frequent at the beginning of relationships which later turn violent, and that such behavior is most often observed in men. Delusional jealousy can be a stand-alone disorder, but is often comorbid with other conditions such as paranoid schizophrenia, alcoholism, drug addiction, and neurodegenerative diseases such as multiple sclerosis, Alzheimer's, Parkinson's or, in certain cases, results from brain trauma, stroke, or a brain tumor.

Couvade Syndrome or Sympathetic Pregnancy

This is again a condition which mainly affects men and in which a (male) partner displays some of the same symptoms and behavior as their expectant (female) partner. Symptoms can include minor weight gain, changes in appetite, altered hormone levels, morning nausea, heartburn, abdominal pain, bloating, a bulging stomach, swollen hands and ankles, nosebleeds, acne, diarrhea or constipation, breast growth, urinary or genital irritations, leg cramps, backaches, toothaches, fatigue, reduced libido, disturbed sleep patterns, anxiety, and depression. Onset is usually during the third gestational month with a secondary rise in the late third trimester. Symptoms can sometimes extend into the postpartum period in the form of postpartum anxiety and depression. With financial worries, health concerns, and ambivalence about parenthood, pregnancy is often felt as an extremely stressful event by both partners. Add a little empathy to the mix, and you may have a perfect recipe for the syndrome. Some psychiatry professionals, however, dismiss it as stand-alone condition and speculate that the symptoms may be the psychosomatic expression of some other psychiatric condition, specifically the combination of anxiety with a mood disorder.

False or Phantom Pregnancy

This condition is similar to the one described above, except that the affected person is typically female. It is thought to be caused by endocrine changes in the body leading to the secretion of hormones that cause physical changes similar to those observed during pregnancy including cessation of periods, morning sickness, tender breasts, an abdominal bulge (thought, in this instance, to be caused by a buildup of gas, fat, urine, and feces), labor pains, increased appetite, and weight gain—all in the absence of an actual fertilization event. Many women even report feeling the baby move or kick. In a false pregnancy, there was no conception and there is no actual baby. Despite this, the strength and duration of symptoms can make a woman, and even those around her, including many health care professionals, believe that she is expecting. The condition is different from simulated pregnancy, where a woman intentionally fakes a pregnancy. Hippocrates gave the first written account of false pregnancy around 300 BC, when he recorded twelve cases of women with the disorder. Some mental health professionals believe that when a woman desperately wants to be pregnant, possibly after experiencing multiple miscarriages, infertility, or because she yearns to get married, she may misinterpret certain changes in her body as a clear sign of pregnancy. Many also think that it may be the manifestation of the psychosomatic symptoms of other psychiatric conditions, the most likely culprit being the combination of anxiety and a mood disorder.

Histrionic Personality Disorder

This condition also primarily affects women. It is a personality disorder characterized by a pattern of excessive attention-seeking behaviors and displays of extreme emotion, usually beginning in early adulthood, including inappropriately seductive behavior and an excessive need for approval. Histrionic people are lively, dramatic, vivacious, enthusiastic, and flirtatious.

121

They are usually high functioning, both socially and professionally. They tend to be socially skilled, but often use those skills to manipulate others into making them the center of attention. They are drama seekers and creators. They often fail to see their own personal situation realistically, instead dramatizing and exaggerating their difficulties. They dress provocatively and use makeup and their physical appearance to draw sexual attention to themselves. They can also use somatic symptoms to garner attention. They seek constant reassurance and approval from those around them and display extreme sensitivity to criticism. They tend to be emotionally unstable and could suffer from a mood disorder comorbidly. Indeed, many tend to get clinically depressed when a relationship ends. This personality disorder bears similarities to two conditions I described previously, namely atypical depression with rejection sensitivity and borderline personality disorder.

Shared or Induced Delusional Disorder

This disorder is also called *folie à deux* or, translated from French, "madness of two." It was included in *DSM-4*, but was ultimately removed in *DSM-5*, most likely because it tends to show up comorbidly with other psychiatric disorders. The condition is diagnosed when two individuals, who live in close proximity and may be otherwise socially or physically isolated, develop the same delusional belief or beliefs, usually of a paranoid type. When both partners in an intimate relationship have a genetic predisposition to psychotic spectrum disorders (schizophrenia, bipolar, borderline, or depression), a stressful life event could trigger the onset of paranoid delusions in one partner. When the affected person is the dominant one in the relationship (typically a male partner), the codependent partner (usually a female) can be induced to develop the very same beliefs, especially if she is also isolated socially. Shared delusional disorder is hard to diagnose because, usually, the afflicted persons do not seek out treatment given that they do not realize that their delusion is abnormal since it comes from someone in a dominant position whom they trust. Besides, the very definition of a delusion is its persistence despite disproving facts. Untreated, it can have serious consequences. Since most delusions are accompanied by fear and paranoia, they increase the amount of stress experienced, leading to increased levels of stress hormones (cortisol, epinephrine, and norepinephrine) which trigger increases in blood pressure, blood sugar levels, heart rate, and breathing rates, which put them at risk of developing cardiovascular disease and diabetes in the long run. Since those hormones also suppress the immune system, they are at elevated risk for a host of infections. The elevated stress levels could also trigger full onset or worsening of any mental disorder the individuals were genetically predisposed to. In a recent study (Nordsletten et al., 2016), researchers analyzed the records of more than 700,000 individuals who had been diagnosed with a wide range of mental disorders. They found that, overall, people with mental disorders were two to three times more likely than the general population to have a romantic partner with any mental disorder. Some disorders showed a greater likelihood of both partners having the same diagnosis. People with schizophrenia, for instance, were seven times more likely to partner with someone else with their condition than the general population, while autistic people are over ten times more likely. People with a mercurial temperament are also known to mate with others who have a predisposition to mood disorders. All this suggests that shared delusional disorder may be more common than current statistics may imply.

7

The Rude Awakening: When Falling in Love Conflicts with Self-Interest

In her book about psychoanalysis, Janet Malcolm provides the following quote from Freud: "Isn't what we mean by 'falling in love' a kind of sickness and craziness, an illusion, a blindness to what the loved person is really like?" I have taken the position in this book that falling in love is, indeed, a type of psychosis. The cocktail of powerful biochemicals released by our brain's love circuitry once it decides to pull the lever of love loosens the grip of our prefrontal cortex on our brain's emotional release center, seriously impairing our ability to make sound judgements, and induces changes in our perception, affect, cognition, and behavior, including hallucinations, delusions, magical thinking, obsessions, and compulsions, reminiscent of psychotic hypomania or certain mental states induced by recreational drugs. The full scale of the consequences from the actions we may have taken while detached from reality only dawns on us in the aftermath of our slow awakening. Our brain's love circuitry is ultimately built to accomplish one, and only one, goal—which is to make reproduction happen. Contrary to myth, love's ultimate goal is not our happiness. Any good parent will tell you about the staggering personal cost to having children (probably along with some of the benefits). Therein lies the conflict with self-interest. In a certain way, falling in love works a bit like a drug-induced self-destruction (not a complete one, ideally, but one that simply inflicts quite a bit of serious damage) for the sake of perpetuating the human species. Hence, those of us who have never had a desire to have children are just as capable of falling in love as those of us who have dreamt of having a family since childhood. Most of us would prefer to have been born to parents who both wanted children in their life—the parents who did not end up feeling trapped in a life they do not enjoy. There is nothing in our brain's love circuitry that would prevent a homosexual person from falling in love with a heterosexual friend, making their life miserable in the process. We may look like a perfectly average female, yet have a male-typical brain (or vice versa) and may, therefore, want nothing to do with our traditionally assigned gender role. Many progressive men and women find out how gendered and traditional their activities become after having a child. Nothing is there to prevent a low-libido person and a high-libido person from falling in love with each other. In fact, being in love would temporarily bring their libido closer to an equilibrium. And, in the heat of passion, they may very well decide to marry each other, however ill-advised many marriage counselors would think such a union to be. Indeed, sex and finances are the two biggest reasons why couples are known to quarrel. In chapter 3, we learned that when and with whom we fall in love is unconsciously determined for us based on evolution. Our ancestral checklist for a long-term mate is rather sketchy when one thinks about it, consisting mainly of checking for youth, beauty, and signs of sexual fidelity in a female mate and for socioeconomic status and signs of commitment and generosity in a male mate. Ancestrally, for both men and women, strangers from our own ethnic group have also been preferred as long-term mates in order to strike the right balance between inbreeding and outbreeding (and perhaps because life was mostly constrained to the confines of a small village for much of human history).

How about common interests? How about intelligence? How about compatibility of life goals? How about commonality of beliefs? How about mental stability and health? How about being roughly in the same life stage? It is not that uncommon to see a forty-something and established male professional marrying a twenty-something college student. It is only when conflicting life goals start to stretch the union towards a breaking point that the mistake becomes apparent. Men's brains have been wired to look for youth as the number one item in potential mates, as a gauge of fertility. Hence, it has traditionally been a major factor in female-female reproductive competition. In fact, until the twentieth century, most twenty-something unmarried girls would have been called "old maids" in much of the world. The view still holds in many parts of the underdeveloped world, where young girls will drop out of high school in droves in order to get married, lest they miss the marriage train altogether. Evolution does not seem to have cared the least bit about women's need for self-development, let alone self-actualization of any sort.

Moreover, if we are most likely to fall in love when we are in an emotionally aroused situation (specifically one resembling the damsel-in-distress archetype) rather than when we are calm and composed, then we are especially prone to making serious mistakes. Just think about the complexity of the problem that our brain's love circuitry is trying to solve as it tries to find a DNA that best meshes with ours. In the language of mathematics, it is akin to having to find one of the equilibrium points to a long list of coupled differential equations with many variables given one set of initial conditions picked from a list of innumerable possible ones. Any smart mathematician would walk away from such a problem and concentrate their efforts on simpler systems. One reason for the observed differences in the degree of happiness and bonding over long periods of time between couples is quite likely the brain differences between individuals. Some brains are ultimately better built to handle complexity than others. Hence, they can make better choices—not just regarding reproductive or romantic partners, but in all areas of life including social, financial, and professional. The damsel-in-distress archetype represents a scenario in which the conflict between survival and reproduction is minimized for both sexes. Thus, women are most likely to fall in love when their brain determines reproducing to be their best chance at surviving (by "exploiting" the economic resources of the prospective mate in times of socioeconomic duress) and men, when reproducing hurts their effort at self-survival the least (by "exploiting" the biological resources of a prospective mate in times of socioeconomic success).

If the constraints we are facing in such circumstances are severe enough, something determined by both the intensity and duration of stress, the lowest-order check may be deemed good enough by our overwhelmed brain. Remember, the lowest-order check simply consists in looking for youth in women and for socioeconomic status in men. It is worthwhile to note that, besides being a gauge of fertility, relative youth of the female is also likely to maximize the difference in socioeconomic status between her and her male suitor, increasing the probability that she will yield to impregnation. This may explain why most women in preindustrialized societies will be married off before reaching the age of twenty, while female fertility is known to peak around age twenty-five. All other checks are simply skipped before we find ourselves quickly blinded by the flood of biochemicals that our brain's love circuitry releases, turning falling in love into a perfect recipe for making very ill-advised, yet life-altering, decisions. We do this because intensely stressful situations evoke our fight-or-flight response. The higher the intensity of the stress and the longer its duration, the stronger our brain's response, the faster its decision making, and the shorter the time horizon for any planning. Our stress response was evolved in times where we regularly faced life-or-death types of short-term stress—as in facing a hyena alone in the middle of the Serengeti. If we could not manage to survive right then, in that very moment, we would not have been present to see tomorrow. Therefore, the only relevant time frame is this very instant; the future does not matter. This is also the reason why millions of retirees will sell out of any stock they may own in their retirement portfolio at a 30% loss in the very midst of a selling stampede during a recessionary period. It is only when tomorrow does come (about a year or two down the road) and they realize that they are still alive and may well remain alive for the next twenty or thirty years that they deeply regret having allowed a vast amount of their wealth to readily be transferred to those who remained calm and got their 30% discount, only to recover the original valuations and go even further up as soon as the sun started to shine back on the economy. In the context of love, it is only when this biological chain reaction ends several years down the road that we slowly wake up and start to see the person we reproduced with in the way they truly are—which is really the way they had always been. By then, if nature did get its way and reproduction has happened, we find ourselves tied to them in a way that no divorce can ever break.

When it comes to selecting an optimal intimate partner, the use of dating websites and applications is by far superior for finding our match than reliance on our evolved mechanism of falling in love. Popular dating websites offer a much wider pool of candidates than the pool made up of those we know. We are fully in touch with reality when we go about browsing profiles, rather than drugged-up and psychotic as we are when in love. Each profile contains far more information than we typically know of acquaintances, so that we do not have to make up anything. Approaching a stranger via an electronic message is a method most people feel rather comfortable with when compared to facing the prospect of meeting someone in person. Online rejection is easily forgotten. By the time we meet the person of our choice for a meal or drink at a restaurant or bar, we know a great deal about them, and they about us—not just from reading online profiles, but also from the email exchange and phone calls we probably had with them beforehand. The first meeting is an opportunity to gain more information with consciousness and purpose, with no obligation to ever meet again. Intimacy builds up over time with someone we connected with as we go about dating them further and having sex with them. Nothing in us falsely compels us to remain with someone we no longer find suitable following a few dates. Nothing in us blinds us to their defects. Building a relationship with someone who is truly a good match to us is easy and natural. Each encounter is enjoyable. Each future date brings about excitement. Sex feels great. Their company feels warm and stimulating. Attachment naturally forms over time. Research backs up the positive experiences that many people have had with online dating. A study of a nationally representative sample of 19,131 respondents who married between 2005 and 2012 revealed that more than one-third of marriages in America now begin online (Cacioppo et al., 2013). Results also indicated that of the continuing marriages, those in which respondents met their spouse online were rated as more satisfying than marriages that began in an offline meeting. The authors hypothesize that the greater amount of self-disclosures and affiliations when strangers meet online compared to face-to-face may explain these promising statistics. In addition, analyses of breakups indicated that marriages that began in an online meeting were 25% less likely to end in separation or divorce than marriages that began in an offline venue. The researchers suggested that the greater pool of candidates online may allow users to be more selective in their choice than the thinner traditional pool of those around us, resulting in better-matched couples.

The following three excerpts (with minor edits) from a Facebook page called "I regret having children" with anonymous contributors are helpful in both illustrating and understanding the discussion of the preceding paragraphs and opening up one on the important topic of parenthood.

Excerpt #1

"I'm now thirty-nine, and after thinking I'd never have children, I find myself a single mother of a four-and-a-half-year-old boy. My son is lovely. He's kind, funny, and sweet. I don't understand why I can't embrace being a mother, and let go of the old me and the life I had before having him? Does anyone else feel like this too? I genuinely can't remember the last time I felt truly happy or free. I used to have the best life: I had awesome friends, I was out every night, and had a career I adored. Whilst I've managed to keep my career (thank goodness!), I hardly ever see my friends, I can almost never go out, and I'm depressed and so fed up that I'm not sure I can put it into words . . . I don't consider myself to be a selfish person; all I want is to be able to live by my own time, and do what I want, when I want. I didn't find out I was pregnant until I was almost three months along. I was engulfed in postpartum depression for the first few years . . . Has anyone been in my position and managed to come out of it—I mean to truly accept their life changes and to move forward and be happy?"

Excerpt #2

"I'm twenty-two, I have two young boys, and I'm a military wife. I haven't

seen my husband in months and he won't be home for a while. My youngest
son is really needy and the minute I put him down he screams, so I get
nothing accomplished, or I get it accomplished but he screams the whole
time. I get judged constantly by my family. Nothing is ever good enough.
They complain that my baby is spoiled, but then complain when I let him
fuss so I can get errands done. Lately, I've been missing our life before we
had children. I feel so bad for my husband because I'm a wreck. I love my
children, but sometimes I regret having them. I see other moms who look
so happy and I feel like something is wrong with me because I don't feel
that way some days. I'm so stressed out that my patience is short and I get
overwhelmed. I just hope I'm not the only one who thinks having children
isn't that great."

Excerpt #3

"I always knew I didn't want children. In fact, I became well known for it
amongst family and friends as I've never been afraid to have an opinion and
voice it. But when I met the love of my life at the age of thirty-one, and he was
very open about wanting a family eventually, I forced myself to reconsider.
Since I couldn't bear the thought of losing him, given how dearly I loved him,
I relented.

Here I am, several years later, with a three-year-old and a four-month-old.
This is what I've learned:

1. If you've not felt some kind of deep, primal urge to have kids, don't do
   it. No amount of love for your partner will make it the right decision.
2. Yes, if I could rewind time, I'd try much harder to convince him to not
   bother me about it rather than giving in so quickly.
3. Motherhood is really hard. It's pure hard graft and it doesn't get easier,
   it just changes. If you want something to care for, get a dog. At least
   you can rehome it if it doesn't work out . . .

4. If you value your independence, time to yourself, any kind of career or hobby, a social life, a non-ruined body, and not feeling like every day is (the movie) Groundhog Day, don't have kids.

5. I may not regret it as much when they move out eventually—please, dear God, make it happen fast—but I am not sure the relief will make up for the eighteen-plus years of lost life before that.

6. Just listen to your intuition, not to anyone else."

On June 9, 1980, while freebasing cocaine and drinking 151-proof rum, thirty-nine-year-old stand-up comedian Richard Pryor entered a drug-induced psychosis during which he poured rum over his body and set himself on fire. He then left his house and ran down his street in Northridge, California, until subdued by police. Despite facing survival odds of one in three, he managed to live on and continued to do stand-up comedy. While in love at age thirty-one, the woman responsible for Excerpt #3 above entered an endogenous-drug-induced psychosis during which she engaged in the self-destructive act of making two children, entirely ignoring the fact that she had always been strongly averse to the idea of having children. She does sound like someone who experienced a rather rude awakening from the intoxicating effect of falling in love. I wonder, in the end, who did more damage to their life.

There are many other similar postings on the same Facebook page, but you get the idea. This is not to say that the end result of falling in love is necessarily doom and gloom. Many women want nothing more in their lives than the ability to make lots of children and stay with them all day, every day, to raise them. Many men want nothing more in their lives than being able to play the role of the provider for such a family. When, fortuitously, the person we fell in love with is not too different from the ideal representation we projected on them while in love, we may, indeed, find our near-perfect match. When we always wanted to have children and get to have them with our near-perfect match, and discover, after having them, that we do love parenting, then being in love does result in a feeling of fulfillment. Our happiness or unhappiness are, however, merely collateral outcomes of a process designed with the sole purpose of perpetuating the human species, not its end goals. In a recent study (Margolis and Myrskylä, 2015), researchers followed 2,016 Germans from three years before the birth of their first child until at least two years post-birth. Participants were asked to rate their happiness on a scale of 0 (completely dissatisfied) to 10 (completely satisfied) in response to the question, "How satisfied are you with your life, all things considered?" They found that most people in their study started out at high levels of happiness when they decided to have their first child. In the year preceding the birth, their life satisfaction went even further up, presumably due to successful pregnancy and anticipation of the baby. After the birth, however, while about 28% of respondents remained at roughly the same level of happiness, the remaining 72% reported decreased happiness, with 37% reporting a one-unit drop, 19% a two-unit drop, and 17% a three-unit drop. To put things in perspective, previous studies have quantified the impact of other major life events on the same happiness scale in the following way: divorce causes a 0.6 happiness-unit drop, unemployment a one-unit drop, and the death of a partner a one-unit drop (Cha, 2015). New parents cite three major causes of decreased happiness: (1) complications during pregnancy and labor; (2) health issues before and after birth; (3) the generally exhausting and physically taxing task of caring for the baby. The researchers also concluded that a drop in well-being surrounding first birth predicts a decreased likelihood of having another child. The association was found to be particularly strong for older parents and those with higher education. In November 1975, advice columnist Ann Landers asked her readers, "If you had to do it over again, would you have children?" For several months, she tabulated responses from nearly 10,000 parents' handwritten postcards before revealing the results in the June 1976 issue of *Good Housekeeping*. To Landers's surprise, 70% of respondents said they would not. Most scientific studies generally find happiness levels of parents—averaged over large samples—to decrease with the birth of the first child and to only increase back up meaningfully when their nest becomes empty. Social scientists have noticed, however, that an overwhelming majority of parents would tell their children that parenthood made them happy, regardless of how they may have actually felt about it. This would make sense from an evolutionary perspective since their only chance at passing their genes into a third generation entirely rests on seeing their children reproduce. Otherwise, their own reproductive sweat and toil will be wasted away in an evolutionary dead end.

With no direct experience before having children, people are uncertain about what pregnancy and childrearing are really like. When the price of parenthood is finally revealed after the birth of a first child, it is sometimes felt to be so unutterably high that postpartum depression ensues. Depression—just like many psychiatric conditions—is a social strategy we have evolved to send a genuine signal of distress and a cry for help to those around us. It is a threshold-dependent mental state, which some of us are more readily prone to resort to than others by virtue of our genetic makeup. The young mother afflicted by postpartum depression is overwhelmed by the heavy toll of pregnancy, birthing, and childcare. It is estimated that clinical depression affects 10% to 15% of adult women in the postpartum period (Ko et al., 2017). Paternal postpartum depression is a phenomenon with high comorbidity with maternal postpartum depression. It was estimated to occur in 4% of fathers in a recent large-scale study (Davé et el., 2010). In many cases, signs of depression are present starting with the first trimester of pregnancy. Parents suffering from postpartum depression report difficulty with the demands of day-to-day functioning and caring for their baby, in addition to problems bonding with their child. Marital discord is common. Parents also endorse fantasies or worries that they may in some way inadvertently harm their infant—for example, by dropping the baby in a time of exhaustion or frustration or even ignoring or injuring the baby (Kim and Swain, 2007). Thinking about it factually, the prospect of losing sleep in order to help someone with the basic activities of daily living is not a particularly appealing one, especially when such demands are made on a round-the-clock and uncompensated basis. As a test, high-school teachers could ask their students to raise their hands if their dream job is to become a nursing home aide, and then tabulate the statistics. Nursing home aides have limited work hours plus salary and benefits, by the way. When the incapacitated person is a baby, or more specifically, our own progeny, the hormones of our attachment system come to help: the female brain is hit by an extreme dose of oxytocin in the last moments of parturition, both to help her forget the pain of pregnancy and labor and to induce bonding with the baby by forging a permanent link between the birth and pleasure in her brain, and then again with smaller doses during episodes of lactation or simple skin contact with the baby to reinforce that link. Remember, oxytocin is the endogenous biochemical equivalent of heroin: it relieves pain, induces mild euphoria, and creates dependency, which is hard to eliminate due to its extremely painful withdrawal symptoms. Some level of psychosis has got to be a part of the effect of this potent biochemical to keep 85% of new mothers relatively enthusiastic about their new day-to-day tasks. It is because this hormone is seriously lacking—in relative terms—in the case of caring for nursing home patients that most people would not rate the job as a pleasant one. Depressed parents may well represent those whose brain has failed to render sufficiently delusional.

Anyone doubting the overwhelming power of our evolved process for falling in love has never experienced it. Think about it. Men have always been free to turn their backs on reproduction, yet most of them have ended up signing up for it. Our presence today attests to that. Far back in humanity's rude past, they routinely risked death in physical male-male confrontations for the mere chance to reproduce. Later, they still risked their lives to acquire social status in hunting or war for the main purpose of having the right to impregnate as many fertile women as they could. The prospect of toiling their lives away in order to feed a large family did not hold most of them back either. Women used to routinely die in childbirth. They lived deplorable lives, going from the misery of one pregnancy to that of the next until they either perished or were rescued by menopause. Here is a short list of the torments that pregnancy can inflict on a woman besides killing her: constipation, intense nausea and violent vomiting (sometimes happening any moment of the day, every day, for the entire duration of the pregnancy), doubling of blood volume resulting in swollen ankles and feet, swollen face and nose, swollen and purple-colored vulva, nosebleeds, hemorrhoids, carpal tunnel syndrome, vaginal varicose veins, and bloody gums, constant sweating, foul smell, acne, new skin tags, bigger moles, leg cramps, increased risk for bacterial vaginosis (and associated fishy vaginal smell), yeast infections (and associated burning crotch), and trichomoniasis, passing gas frequently and uncontrollably, loss of libido (sometimes permanently), uncomfortable or painful sex (sometimes forever), restless leg syndrome, muscle spasms, sciatic nerve pain, bruised ribs, enormous nipples, irritation of the uterus causing painful contractions throughout pregnancy, round ligament pain, acid reflux disease, shortness of breath while walking, pregnancy brain (described by some women as becoming foggy and forgetful—research has shown that some of the brain modifications induced by pregnancy may be permanent), depression and anxiety (both during pregnancy and postpartum), postpartum PTSD, postpartum psychosis, weight gain, stretch marks, saggy breasts, and perineal lacerations. None of this ever stopped most women. Surely, many were abandoned by their male partner and left to fend for themselves and their children alone. Yet, most of them still did not opt out of reproduction. The prospect of toiling their lives away in the care of babies, toddlers, and children did not deter the reproductive need. Granted, they only gained the luxury of reproductive choice roughly mid-twentieth century. At the time of writing this book, women have been free to say "thanks, but no" to reproduction for several decades, yet most of them have continued to sign up for it. At roughly 10% today, despite what one might think, the percentage of childfree women roughly equals that of men.

Some may fear that talking this openly about the challenges of childbearing and rearing might dissuade young people from conceiving one day. The fear is unfounded. Just like praising homosexuality will not persuade a group of firmly heterosexual men and women to change their primary sexual orientation, having a frank discussion about the wide range of feelings reproduction may evoke among us will neither persuade nor dissuade anyone from their deep desire to become or not become a parent. Many have affirmed that there is no instinct or primal urge to have children. I think the contrary to be true. Some people have dreamt of having children since childhood. For many among those, neither an eventual bitter divorce nor having children with severe mental or physical disabilities have changed their view that becoming a parent was worth it. At the opposite end, some have felt nothing but repulsion at the idea of pregnancy and have spent their reproductive years avoiding both marriage and conception. No amount of life experience or learning has turned their feelings around. Most people are somewhere between these two extremes. The urge to become a parent appears to lie on a continuum just like the rest of our psychological traits. For those born with little to no desire to have children, falling in love can sometimes act as a catchall if necessity has failed to enlist them into parenthood. The tragedy of falling in love is threefold: (1) It is entirely independent from our deep feelings towards parenthood, sometimes leaving those who were ambiguous about it with the sense of being trapped and regretful after having children; (2) More often than not, it is not reciprocal—the misery of unrequited love is all too common, hence our fascination with tales of reciprocal love; (3) Many times, it results in pairing us up with wholly inappropriate partners—our modern wish list for a long-term mate tends to be a lot more elaborate than our sketchy ancestral wish list. Besides, being in love makes us fabricate an imaginary being who has all the qualities of our ideal long-term mate, and then become absolutely convinced that the fabricated being and the actual person are one and the same, leading to a rather rude awakening when we slowly come out of our love-induced psychosis.

Falling in love is a process patiently sculpted over time by the chisel of necessity. One does not need to dig too deep into an intimate relationship to uncover some form of dependency lurking beneath the virtuous lacquer of love. We are a highly specialized species in which men have heavily depended on women for reproduction while women have extensively relied on men for survival. As the human baby started to be delivered earlier and earlier, hence in a more and more vulnerable state, and as childhood lengthened—both events being direct results of our ever bigger and ever more complex brains—monogamy over a more extended period of time became a necessity. Those who were able to stay together for longer benefited from the pooling of resources that monogamy allowed. Both they and their children fared better than those left alone to fend for themselves and their progeny. Why a man should choose to stay with one woman instead of running away after impregnating her has been a puzzling question to many social scientists. But should the woman choose to abandon the baby, or should the two them end up starving from the handicap of pregnancy and its aftermath, the man will not have achieved much in the way of reproduction. Why a woman should not take the sexual invites of the attractive young man next door instead of remaining loyal to her aging husband with middling looks is the result of a risk-reward equation—she and her children may lose access to the resources and protection of her current mate. The allure of achieving biological diversification via extra-pair copulation is, of course, the great enemy of monogamy. This is where our brains come to help by soothing us with the drugs of our attachment system—oxytocin, vasopressin, and the related endorphins. These drugs have the necessary psychotic effect of blinding us to the defects of our loved ones to protect our mental representations of them from our own critical judgement for the sake of preserving unity. The decision to become a parent for the first time is a daunting one given the enormity of the biological and economic resources one is bound to give away to the future child, not to mention the loss of our personal freedom. And that's where our brains come to help once again with the drugs of our attraction system—dopamine, serotonin, norepinephrine, NGF, and testosterone, among others. Suddenly, our middling future mate merges with his or her diamond-studded, idealized, and dazzling mental representation. And in the heat of our drug-induced euphoria—which shares quite a bit of commonality with hypomania—we both jump off that cliff happily, only to slowly awaken after the baby arrives.

Among the many delusions we tend to hold while in love is the unshakeable belief that we are meant to remain in this mental state forever, that we will be in love with our beloved for eternity, and that nothing can ever happen to stop the incredible rush we are feeling. In reality, the growing level of euphoria and rush that overtakes us in stage two of falling in love is only meant to create a sense of urgency in us so that we may take reproductive action and procreate. The rush is bound to stop once pregnancy has successfully taken place, meaning once nature has accomplished its goal. Stage two of falling in love, the stage of maximum passion, is a biologically expensive process where our body is maintained in the fight-or-flight response state for an extended duration, leading to increased levels of stress hormones (cortisol, epinephrine, and norepinephrine), triggering increases in blood pressure, blood sugar levels, heart rate, and breathing rate which, in turn, put us at risk of developing cardiovascular disease and diabetes. Important nutrients are diverted away from bodily functions deemed nonessential such as building muscle and bones, supporting our digestive system, and maintaining the integrity of our immune system, making us vulnerable to a host of infections. Besides, having to depend on two psychotic parents is not going to help the baby survive for very long. The process cannot, however, be terminated in its entirety with a pregnancy event given the nine months needed for readiness to delivery, how vulnerable a human baby is, and the extended duration of infancy. Thus, the abrupt drop in euphoria at successful pregnancy is followed by a slow exponential decay of the remaining feeling of pleasure over a four-to-five-year period to allow for time to accomplish those tasks. It is only then that we come to the realization that the rush was truly not meant to last forever. If stage two of falling in love lasts for two to three years, at most, and stage four for four to five years, given that stage one and three only take a fraction of a second, is all we are left with the "seven-year itch" to part ways?

If 50% of all first marriages in the US end up in divorce, the remaining 50% stay intact for a lifetime, implying the presence of factors beyond the biochemistry of love for long-term relationship survival. Research shows, for instance, that individuals who score low on the personality trait of neuroticism and high on that of conscientiousness tend to have relationships which don't just last longer, but are also more harmonious and satisfying. Scoring high on neuroticism increases an individual's chances for a number of negative life outcomes, including relationship discord, divorce, professional instability, substance abuse, and the development of anxiety and mood disorders once a certain stress threshold is crossed. Scoring high on conscientiousness, on the other hand, is associated with a number of positive life outcomes extending beyond intimate relationship strength into all aspects of both the social and professional spheres. One can easily fathom how scoring high on emotional intelligence (EQ), which is the ability to decode the emotions and motivations of others, can positively impact the quality of a relationship—but only if one is also capable of compassion. The dark triad personalities of sociopathy, narcissism, and Machiavellianism—which tend to principally manifest themselves in men and are characterized by a normal or higher EQ, but a natural inability to feel negative emotions such as sadness, fear, and anxiety, hence a serious deficiency in compassion—tend to lead to callous behaviors and would be expected to have the opposite impact on relationship quality. Severely deficient EQ characterizes autism spectrum disorders (principally affecting men) and translate into difficulty with social interaction and communication in general. Lack of maturity, excessive self-doubt, and inadequate self-reliance, perhaps due to excessive anxiety (which tends to primarily impact women) translate into clingy and overly dependent behaviors and can also substantially lower the quality of a relationship. If our personality is fixed-in-stone by the time we are twenty-five years old, are some of us destined to live through a string of dissatisfying and broken relationships? In the same way, there is a positive correlation between any of the three variables consisting of IQ, educational attainment, and length of marriage. A healthy libido in both partners can immensely contribute to relationship satisfaction and survival. Indeed, sexual activity has the power to bond two individuals in a way that no friendship or platonic feelings ever can. There is also a positive correlation between any two of the three variables consisting in libido, mental health, and physical health. Hence, both mental and physical health are conducive to more satisfying relationships. In most cases, family dysfunction stems from the severe mental or physical disability of one or more members. Most, if not all, the factors I just mentioned are, to a large extent, preordained by our DNA at the moment of conception which, once again, begs the question of whether some of us are genetically doomed to a life of instability and frustration. Besides the assistance that modern medicine can provide and the fact that our DNA can, to a certain extent, be modified epigenetically, we are intelligent animals who can exert a degree of care into choosing some of the elements in our environment, including those with whom we choose to associate. This is where the notion of complementarity can come to help. If we choose our partner astutely enough, our strengths can complement their weaknesses and vice versa, leading to a balanced and stable relationship. For certain traits, such as sexual desire, similarity is of prime importance. Each of us constantly, albeit unconsciously, weights the positives of our intimate relationship versus its negatives. As long as the give-and-take that is inherent to all relationships remains roughly balanced and the sum total of all positives slightly exceeds the sum total of all negatives, the bond will remain intact. But the moment that one partner starts to feel that they are giving a lot more than that they are getting out of a union, cracks will start to appear in its foundation. Besides, given our ever-increasing lifespans, it may be a bit too much to expect from anyone that they remain attached to us for sixty or seventy years straight. Our maturity and our ability to stand on our own legs and be independent are ever more important. These same traits are also those that make us a more attractive mate, by the way, regardless of our sex. There are many people out there with whom each of us can bond. Serial monogamy can involve a string of happy relationships of varying lengths across our lifespan. Human evolution lags behind medical and technological progress. Our evolved process for falling in love cannot possibly catch up with modernity in any of our lifetimes. It remains a task for our intellects and imaginations to make up for the shortfall.

The fact that we tend to fall in love in a situation akin to the damsel-in-distress archetype is not trivial. Such a circumstance is one where the conflict between reproduction and self-interest is minimized, perhaps even reversed, at least temporarily. Falling in love in times of socioeconomic distress has ancestrally allowed women to survive and get help at a time of dire need; meanwhile, falling in love in times of socioeconomic success has ancestrally allowed men to share their material resources at a time when it would have hurt them the least, and thus help them to reproduce. Until recently, male provisioning in the context of reproduction was the principal means of livelihood for the great majority of women. For men, reproduction allowed access to strategic male-male coalitions or the formation of a new kin-based coalition serving to enhance their social status and resource acquisition abilities. Children used to be the only retirement plan for both men and women. Ancestrally, children took away quite a bit from their parents but, at least on average, they gave quite a bit back to them over time. Modernity has empowered individuals. The notion of relying on kin and on close neighbors in situations of economic distress has been replaced by the welfare state. Country-level judicial systems have abolished the rule of tribal law. The educational system has displaced the blind following of religion and traditions. The medical system comes to help during health crises. Corporations provide their employees with the resources they need today and in retirement. Many institutions have been built to aid in old age, from social security, Medicare, and Medicaid, to hospitals, retirement communities, assisted-living facilities, nursing homes, or community-based organizations. We are no longer geographically restricted to the village we might have been born into or to the small community made up of its population. We travel far with ease, live in large metropolitan areas, and are connected globally thanks to the internet. We are all relieved from the tyranny of having to guard our reputations within a small community of people—we can move away and start afresh. We no longer have to bear the presence of toxic people in our life forever—we can divorce, break up relationships, move away, or obtain restraining orders and police protection. The twentieth century has done more to transform women's lives and empower them than any other century in the entire human history and prehistory. For women, the invention of reliable birth control in their own hands, giving them reproductive choice and the ability to control their own bodies, ranks at the same level (if not higher) than the control of fire in terms of its transformative power. Thanks to it, men—just like women—get to enjoy more casual sex. Men no longer have to support large families on their own. Women are here to help. With greatly diminished infant and child mortality, there is no need for relentless conception either. Modernity does, however, leave reproduction in a quandary. For many men and women who have achieved upper-middle-class or higher social status, having children has become mainly a cost with little benefit besides helping to perpetuate their genes and social and moral satisfaction. Governments still have an interest, however, in replenishing their populations. As imagined by Aldous Huxley in *Brave New World*, children being made in laboratories and raised by institutions is becoming closer to reality. What role would be left for falling in love in such a world? Will it become a maladaptive mechanism destined to gradual disuse and regression over time?

It seems paradoxical that we should have an inborn mechanism which sometimes compels us to engage in actions that are entirely detrimental to our biological selves and our self-interests. In retrospect, rejection of our unrequited love is sometimes the blessing-in-disguise that helped to save us from self-inflicted devastation. The notion of free will advocated by many philosophers seems utterly absurd considering the existence of our brain's love circuitry and the modern discovery of our DNA-based preprogramming. Free fall is always governed by the laws of physics. Free will is always constrained by both biology and environment. If walking is a constant fight against the gravitational force, exerting our self-interest is sometimes a fight against our reproductive drive. Often, our capacity to act rationally and exert our willpower seems to go countercurrent with our biochemical drive to act in the interest of the human species. Just like young trees growing out of the stumps of dead ones, humanity regenerates from the ashes of people gone. Reproduction demands self-sacrifice. Any good parent will tell you how much raising a child has taken away from them. Biological determinism is obviously not absolute. If it were, evolution would be barred in its entirety. Our DNA, the biological program that makes us who we are, is an open software which is slowly being modified via activations, deactivations, and mutations as we interact with our environment. If our biochemistry can influence our thoughts, our thinking can, in turn, influence our biological makeup. Our brain's physiology is certainly being modified as a result of our life experiences or the learning of new skills. Along with some of those experiences and skills, some of our thinking may thus be transmissible to the next generation via reproduction. This principle, perhaps, inspired the thinkers of the Far East with the concept of reincarnation. Through reproduction, a small part of us gets to persist beyond our death. Could this also be, in part, the inspiration behind the idea of eternal life after death present in most organized religions? We self-destruct reproductively so that an ever-diminishing part of us may persist into eternity.

Falling in love is a neurochemistry-driven combination process which bundles attraction, idealization, courting, mating, and attachment into one streamlined package in an attempt to ensure successful reproduction, meaning the birthing of a child and the presence of two strangers-turned-into-kin who will stay together long enough to get this child through the first few critical years of their life. Each individual component of this combination package is a stand-alone process. Falling in love is not necessary for attraction to develop. We all know this experientially from having felt sexually attracted to a stranger passing by without becoming obsessed with them in any way and with no idealization or attachment of any kind taking place. Many of us have enjoyed casual sex for a short while with someone and moved on without them in our lives without feeling too much distress about it. Arranged marriages used to be the norm for much of human history and remain the norm in large swaths of the world. In most of these marriages, the bride and groom have never even held hands and have barely spoken to each other before the marriage ceremony, yet most end up developing attachment to each other over time. Thus, falling in love is not necessary for attachment to develop either. Why then have we evolved it at all? Well, before we became thinking creatures, we were mindless ones driven into action instinctively, meaning via pure biochemistry. As much as we pride ourselves on being the only species capable of thought-driven action, we have kept in us a mechanism which resembles in a lot of ways that which drives two monkeys into conceiving.

A recurring theme of this chapter—and of this book—is the idea that psychosis may not just be a part of our generic makeup, but may also play a crucial role in perpetuating the species. The fact that most of us have fallen in love at least once in our lifetime is proof that most of us are endowed with brain chemistry capable of going off-kilter in certain situations, triggering psychosis. That the core stage of falling in love should resemble hypomania points to the essential role of this mental state in inducing us to take reproductive action. Schizophrenia may well represent the most extreme form of the familiar emotion of jealousy, which we evolved to help in mate guarding. The fact that one in four women will experience depression in her lifetime may indicate a useful evolutionary role for the condition or, at the very least, for a mild version of it. Indeed, both pregnancy and the care of infants have traditionally required women to slow down considerably and lose enough interest in their surroundings to sit still and let time go by. Otherwise stated, a mild dose of what we generally label as mental illness may simply be our norm. Any trait we have evolved has necessarily been useful in some way. Diversity of individuals is generally good for a species to survive in changing environmental conditions. Typically, it is not so much in the presence or absence of a trait but in the gradation of each trait which humanity has seen worth keeping that we differ among ourselves. Just as some of us have much greater mathematical ability than some others, some of us have a much lower threshold to become psychotic or manic or depressed than others. This latter tendency would also make us more prone to falling in love more often and more easily. The traits that characterize psychiatric conditions represent the tail ends of bell-curve-shaped distributions. Being at the extreme end of the distribution of a trait can be a good or a bad thing depending upon the environment and circumstances we face. One can argue that those endowed with extreme versions of certain traits may be more environment-selective than those who have inherited more moderate flavors of the same traits.

It is only when placed in the wrong environment that those extreme traits can manifest themselves as psychiatric conditions. As an example, consider the case of two young women who both exhibit strong separation anxiety and a proneness to clinical depression whenever a critical relationship ends. Let's name them Woman A and Woman B. Woman A meets an empathetic and caring mate who values lifelong relationships at any cost. Woman B meets an antisocial narcissist who likes to exploit others for his sole benefit. One can easily see that Woman B is much more likely to seek treatment at some point in her future than Woman A, despite the fact that both are quite similar in the traits we considered. In fact, Woman B may end up being labeled mentally ill, while Woman A may be regarded as perfectly normal. This example helps to illustrate the veracity of the fundamental thesis behind the diathesis-stress model of mental disorders which asserts that an individual will only develop a clinically diagnosable disorder if the combination of their predisposition or vulnerability (coded in their DNA) and the stress (within their environment) exceeds a threshold. Many otherwise talented people, who also happen to be endowed with a rather low threshold for developing psychosis spectrum disorders, have achieved tremendous professional success by directing the latter tendency into creative fields. For instance, the great physicist Isaac Newton is believed to have had a number of what, today, would be deemed severe versions of psychosis spectrum disorders. Had he followed the wishes of his mother and become a farmer instead, his psychotic tendencies may have hurt a lot more than helped. In that sense, stating that being in love is being in a psychotic state does not necessarily carry a pejorative connotation. The detachment from reality will help, not hurt, someone who has a natural, deeply rooted desire to become a parent, assuming that their actual and idealized mates are not too different from each other in the end. Besides, falling in love is the most psychologically rewarding experience possible for a human being, a mental state of pure delight, and our natural reward for all the work we put into childbearing and rearing. Regardless of the subject matter, knowledge is better than ignorance. My wish for my readers is that they may strive to better know their biological selves and equip themselves with enough knowledge to make conscious choices that will benefit them more than they will hurt.

Author's Note

Thank you for reading *Of Lovers, Lonely Hearts, and the Psychotic Spell Called Falling in Love*. It has been my pleasure to write it.

If you enjoyed reading the book, please consider sharing your thoughts by posting a review on https://www.amazon.com.

I am happy to answer your questions via the "ask the author" feature on my author page at https://www.goodreads.com.

Please feel free to follow me on http://www.amazon.com/author/talzoya.

Talzoya

REFERENCES AND SUGGESTIONS FOR FURTHER READING

Throughout the book, I made extensive use of articles available on Wikipedia. In addition, the references that follow served as valuable sources of information and helped to shape the thoughts which I presented in this book.

BOOKS

Baron-Cohen, Simon. *The Essential Difference: Men, Women and the Extreme Male Brain.* London: Allen Lane, 2003.

Baumeister, Roy F. *Is There Anything Good About Men? How Cultures Flourish by Exploiting Men.* United States: Oxford University Press, USA, 2010.

Brizendine, Louann. *The Female Brain.* United States: Potter/Ten Speed/Harmony/Rodale, 2007.

Buss, David M. *The Evolution of Desire: Strategies Of Human Mating.* United States: Basic Books, 1994.

Darwin, Charles. *The Descent of Man, and Selection in Relation to Sex.* United Kingdom: D. Appleton, 1871.

Fisher, Helen. *Why We Love: The Nature and Chemistry of Romantic Love.* United States: Henry Holt and Company, 2005.

Flexner, James Thomas. *Washington: The Indispensable Man*. United States: Open Road Media, 2017.

Huxley, Aldous. *Brave New World*. United Kingdom: RosettaBooks, 2010.

Huxley, Aldous, Michael Horowitz, Cynthia Palmer. *Moksha: Aldous Huxley's Classic Writings on Psychedelics and the Visionary Experience*. United States: Inner Traditions/Bear, 1999.

Jamison, Kay R. *Touched with Fire: Manic-Depressive Illness and the Artistic Temperament*. United Kingdom: Free Press, 1994.

Kahneman, Daniel. *Thinking, Fast and Slow*. United States: Farrar, Straus and Giroux, 2011.

Liebowitz, Michael R. *The Chemistry of Love*. United States: Berkley Books, 1984.

Malcolm, Janet. *Psychoanalysis: The Impossible Profession*. United States: Knopf Doubleday Publishing Group, 2011.

Money, John. *Love and Love Sickness: The Science of Sex, Gender Difference, and Pair-Bonding*. United Kingdom: Johns Hopkins University Press, 1980.

Plomin, Robert. *Blueprint: How DNA Makes Us Who We Are*. United Kingdom: MIT Press, 2019.

Rako, Susan, Harvey Mazer. *Semrad: The Heart of a Therapist*. United States: iUniverse, 2003.

Shakespeare, William. *As You Like It*. United Kingdom: Simon & Schuster, 2004.

Stendhal. *De L'Amour*. United States: Creative Media Partners, LLC, 2019.

Tallis, Frank. *Love Sick: Love as a Mental Illness*. United Kingdom: Random House, 2009.

Vincent, Lucy. *Comment Devient-On Amoureux?*. France: Editions Odile Jacob, 2004.

Woolf, Virginia, Joanne Trautmann, Nigel Nicolson. *The Letters of Virginia Woolf*. United Kingdom: Hogarth Press, 1975.

PEER-REVIEWED ARTICLES

Bartels, Andreas, and Semir Zeki. "The neural basis of romantic love." *Neuroreport* 11, no. 17 (2000): 3829-3834.

Baumeister, Roy F., and Jean M. Twenge. "Cultural suppression of female sexuality." *Review of General Psychology* 6, no. 2 (2002): 166-203.

Bhatia, Manjeet S., Satish K. Verma, and O. P. Murty. "Suicide notes: psychological and clinical profile." *The International Journal of Psychiatry in Medicine* 36, no. 2 (2006): 163-170.

Breiding, Matthew J., Sharon G. Smith, Kathleen C. Basile, Mikel L. Walters, Jieru Chen, and Melissa T. Merrick. "Prevalence and characteristics of sexual violence, stalking, and intimate partner violence victimization—national intimate partner and sexual violence survey, United States, 2011." *Centers for Disease Control and Prevention, Morbidity and Mortality Weekly Report. Surveillance Summaries (Washington, DC: 2002)* 63, no. 8 (2014): 1-18.

Brown, F., I. Harris, R. Leakey, and A. E. Walker. "Early *Homo erectus* skeleton from West Lake Turkana, Kenya." *Nature* 316, no. 6031 (1985): 788–792.

Brunet, Michel, Franck Guy, David Pilbeam, Hassane Taisso Mackaye, Andossa Likius, Djimdoumalbaye Ahounta, Alain Beauvilain, et al. "A new hominid from the Upper Miocene of Chad, Central Africa." *Nature* 418, no. 6894 (2002): 145-151.

Cacioppo, John T., Stephanie Cacioppo, Gian C. Gonzaga, Elizabeth L. Ogburn, and Tyler J. VanderWeele. "Marital satisfaction and break-ups differ across on-line and off-line meeting venues." *Proceedings of the National Academy of Sciences* 110, no. 25 (2013): 10135-10140.

Cha, A. E. "It turns out parenthood is worse than divorce, unemployment—even the death of a partner." *The Washington Post*, August 11, 2015 at 8:15 a.m. EDT. (Author's note: This reference is a newspaper article and the only non-peer-reviewed publication listed in this section.)

Chrousos, George P. "Stress and disorders of the stress system." *Nature Reviews Endocrinology* 5, no. 7 (2009): 374-381.

Davé, Shreya, Irene Petersen, Lorraine Sherr, and Irwin Nazareth. "Incidence of maternal and paternal depression in primary care: a cohort study using a primary care database." *Archives of Pediatrics & Adolescent Medicine* 164, no.11 (2010): 1038-1044.

Dunsworth, Holly M., Anna G. Warrener, Terrence Deacon, Peter T. Ellison, and Herman Pontzer. "Metabolic hypothesis for human altriciality." *Proceedings of the National Academy of Sciences* 109, no. 38 (2012): 15212-15216.

Emanuele, Enzo, Pierluigi Politi, Marika Bianchi, Piercarlo Minoretti, Marco Bertona, and Diego Geroldi. "Raised plasma nerve growth factor levels associated with early-stage romantic love." *Psychoneuroendocrinology* 31, no. 3 (2006): 288-294.

Finkenauer, Catrin, Olivier Luminet, Lydia Gisle, Abdessadek El-Ahmadi,

Martial Van Der Linden, and Pierre Philippot. "Flashbulb memories and the underlying mechanisms of their formation: toward an emotional-integrative model." *Memory & Cognition* 26, no. 3 (1998): 516-531.

Fisher, Helen E., Lucy L. Brown, Arthur Aron, Greg Strong, and Debra Mashek. "Reward, addiction, and emotion regulation systems associated with rejection in love." *Journal of Neurophysiology* 104 (2010): 51-60.

Frieze, Irene H., and Keith E. Davis. "Introduction to stalking and obsessive behaviors in everyday life: assessments of victims and perpetrators." *Violence and Victims* 15, no. 1 (2000): 3-5.

Geschwind, Norman, and Albert M. Galaburda. "Cerebral lateralization: biological mechanisms, associations, and pathology: I. A hypothesis and a program for research." *Archives of Neurology* 42, no. 5 (1985): 428-459.

Guay, A., J. Jacobson, R. Munarriz, A. Traish, L. Talakoub, F. Quirk, I. Goldstein, and R. Spark. "Serum androgen levels in healthy premenopausal women with and without sexual dysfunction: Part B: Reduced serum androgen levels in healthy premenopausal women with complaints of sexual dysfunction." *International Journal of Impotence Research* 16, no. 2 (2004): 121-129.

Hines, Melissa, Mihaela Constantinescu, and Debra Spencer. "Early androgen exposure and human gender development." *Biology of Sex Differences* 6, no. 1 (2015): 3.

Hoekzema, Elseline, Erika Barba-Müller, Cristina Pozzobon, Marisol Picado, Florencio Lucco, David García-García, Juan Carlos Soliva, et al. "Pregnancy leads to long-lasting changes in human brain structure." *Nature Neuroscience* 20, no. 2 (2017): 287-296.

Ishimoto, Hitoshi, and Robert B. Jaffe. "Development and function of the human fetal adrenal cortex: a key component in the feto-placental unit." *Endocrine Reviews* 32, no. 3 (2011): 317-355.

Kendrick, Keith M. "The neurobiology of social bonds." *Journal of Neuroendocrinology* 16, no. 12 (2004): 1007-1008.

Kim, Pilyoung, and James E. Swain. "Sad dads: paternal postpartum depression." *Psychiatry (Edgmont)* 4, no. 2 (2007): 35-47.

Ko, Jean Y., Karilynn M. Rockhill, Van T. Tong, Brian Morrow, and Sherry L. Farr. "Trends in postpartum depressive symptoms—27 States, 2004, 2008,and 2012." *Centers for Disease Control and Prevention, Morbidity and Mortality Weekly Report* 66, no. 6 (2017): 153–158.

Kranz, Georg S., Wolfgang Wadsak, Ulrike Kaufmann, Markus Savli, Pia Baldinger, Gregor Gryglewski, Daniela Haeusler, et al. "High-dose testosterone treatment increases serotonin transporter binding in transgender people." *Biological Psychiatry* 78, no. 8 (2015): 525-533.

Laumann, Edward O., Anthony Paik, and Raymond C. Rosen. "Sexual dysfunction in the United States: prevalence and predictors." *JAMA* 281, no. 6 (1999): 537-544.

LeVay, Simon. "A difference in hypothalamic structure between heterosexual and homosexual men." *Science* 253, no. 5023 (1991): 1034-1037.

Lester, David, Priscilla Wood, Christopher Williams, and Janet Haines. "Motives for suicide—a study of Australian suicide notes." *Crisis: The Journal of Crisis Intervention and Suicide Prevention* 25, no. 1 (2004): 33-34.

Little, Anthony C., Benedict C. Jones, and Lisa M. DeBruine. "Facial attractiveness: evolutionary based research." *Philosophical Transactions of the Royal Society B: Biological Sciences* 366, no. 1571 (2011): 1638-1659.

Marazziti, Donatella, Hagop S. Akiskal, Alessandra Rossi, and Giovanni B. Cassano. "Alteration of the platelet serotonin transporter in romantic love." *Psychological Medicine* 29, no. 3 (1999): 741-745.

Marazziti, Donatella, and Domenico Canale. "Hormonal changes when falling in love." *Psychoneuroendocrinology* 29, no. 7 (2004): 931-936.

Margolis, Rachel, and Mikko Myrskylä. "Parental well-being surrounding first birth as a determinant of further parity progression." *Demography* 52, no. 4 (2015): 1147-1166.

McCarthy, Margaret M. "Estradiol and the developing brain." *Physiological Reviews* 88, no. 1 (2008): 91-124.

Nordsletten, Ashley E., Henrik Larsson, James J. Crowley, Catarina Almqvist, Paul Lichtenstein, and David Mataix-Cols. "Patterns of nonrandom mating within and across 11 major psychiatric disorders." *JAMA Psychiatry* 73, no. 4 (2016): 354-361.

Pawlowski, Boguslaw. "Variable preferences for sexual dimorphism in height as a strategy for increasing the pool of potential partners in humans." *Proceedings of the Royal Society of London. Series B: Biological Sciences* 270, no. 1516 (2003): 709-712.

Pedersen, Cort A., and Arthur J. Prange. "Induction of maternal behavior in virgin rats after intracerebroventricular administration of oxytocin." *Proceedings of the National Academy of Sciences* 76, no. 12 (1979): 6661-6665.

Petrosky, Emiko, Janet M. Blair, Carter J. Betz, Katherine A. Fowler, Shane P. D. Jack, and Bridget H. Lyons. "Racial and ethnic differences in homicides of adult women and the role of intimate partner violence—United States, 2003–2014." *Centers for Disease Control and Prevention, Morbidity and Mortality Weekly Report* 66, no. 28 (2017): 741-746.

Singh, Devendra. "Adaptive significance of female physical attractiveness:
role of waist-to-hip ratio." *Journal of Personality and Social Psychology* 65, no.
2 (1993): 293-307.

Soma, Kiran K., Nikki M. Rendon, Rudy Boonstra, H. Elliott Albers, and
Gregory E. Demas. "DHEA effects on brain and behavior: insights from
comparative studies of aggression." *The Journal of Steroid Biochemistry and
Molecular Biology* 145 (2015): 261-272.

Van Leengoed, E., E. Kerker, and H. H. Swanson. "Inhibition of post-partum
maternal behaviour in the rat by injecting an oxytocin antagonist into the
cerebral ventricles." *Journal of Endocrinology* 112, no. 2 (1987): 275-282.

Watson, Paul J., and Paul W. Andrews. "Toward a revised evolutionary
adaptationist analysis of depression: the social navigation hypothesis."
Journal of Affective Disorders 72, no. 1 (2002): 1-14.

Wedekind, Claus, Thomas Seebeck, Florence Bettens, and Alexander J.
Paepke. "MHC-dependent mate preferences in humans." *Proceedings of the
Royal Society of London. Series B: Biological Sciences* 260, no. 1359 (1995):
245-249.

www.ingramcontent.com/pod-product-compliance
Lightning Source LLC
Chambersburg PA
CBHW070754290326
41931CB00011BA/2016